M. A. Kadirova
S. S. Rakhimkhodjaev

Production of fabrics and sleeves from high yarn count yarns

M. A. Kadirova
S. S. Rakhimkhodjaev

Production of fabrics and sleeves from high yarn count yarns

On a stnk with a shuttle-gripper

ScienciaScripts

Imprint

Any brand names and product names mentioned in this book are subject to trademark, brand or patent protection and are trademarks or registered trademarks of their respective holders. The use of brand names, product names, common names, trade names, product descriptions etc. even without a particular marking in this work is in no way to be construed to mean that such names may be regarded as unrestricted in respect of trademark and brand protection legislation and could thus be used by anyone.

Cover image: www.ingimage.com

This book is a translation from the original published under ISBN 978-620-7-48430-0.

Publisher:
Sciencia Scripts
is a trademark of
Dodo Books Indian Ocean Ltd. and OmniScriptum S.R.L publishing group

120 High Road, East Finchley, London, N2 9ED, United Kingdom
Str. Armeneasca 28/1, office 1, Chisinau MD-2012, Republic of Moldova, Europe
Managing Directors: Ieva Konstantinova, Victoria Ursu
info@omniscriptum.com

Printed at: see last page
ISBN: 978-620-8-37000-8

Contents

Annotation

The work is devoted to the production of fabrics and sleeves from yarns of high linear density on a shuttle-gripper loom. The shuttle weaving machine is modernised by using the nature of the movement of the reims with the installation of the mechanisms for feeding and withdrawal of the weft to the shuttle grip on the reims. The parameters of the shuttle-gripper are determined. Regularities of movement of the shuttle-gripper in the shed are obtained. The equations of tension of a sinker sliding on a plane, on a circle of stationary and mobile cylinders, taking into account rigidity of a sinker, radius, angle and coefficient of friction are received. It is expedient to use a movable cylinder as a shuttle catch. The values of the angle of friction of the sinker against the gripper depending on the position of the shuttle-gripper in the shed are determined. A stand and a methodology for determining the coefficient of friction of the weft against the shuttle grip depending on the shape, size and condition of the friction surfaces of the shuttle grip have been developed. Increasing the radius of friction of the yarn on the hook grip leads to an increase in the tension of the weft yarn. The coefficient of friction at rest is greater than the coefficient of friction in motion in all variants of rubbing surfaces thread-grip shuttle. The new fabric has a selvedge, the structure of which provides strength and prevents weaving of the extreme warp yarns and does not require additional mechanisms of selvedge formation. In terms of physical and mechanical properties the new fabric is not inferior to the standard fabric. The weft loss for the new fabric increases, which is from 5.6 to 9%, depending on the size of the weft loop in the selvedges. The fabric for production of fire hoses was designed, as a result of which the necessary technological parameters for fabric production were calculated. The new fabric should have water-resistant properties, which can be achieved with the help of special impregnations. The fabric samples were tested for physical and mechanical properties in the laboratory of the certification centre, the characteristics of breaking load, resistance to abrasion, water resistance were investigated. Our fabric has good tensile properties, breathability and water resistance.

Keywords: yarn, warp, weft, fabric, rapport, weave, sleeves, selvedge, parameters, regularity, workmanship, density, breathability, water resistance, properties, tension, friction, shuttle, losses, mechanism.

GENERAL CHARACTERISATION OF WORK

This work is carried out on the following topics: production of fabrics from yarns of high linear density on the machine with shuttle-gripper; design and manufacture of fire hoses with water-resistant properties; and represents the development of a set of basic parts, providing the most complete coverage of data, opening the possibility of effective implementation of the results of work in the production of the textile industry of Uzbekistan in accordance with the state priorities of the economy and industry at present, trends and realities of the market from the current time Intensive socio-economic development of the Republic of Uzbekistan stipulates the necessity of development of new technologies oriented on expansion of the range of consumer goods with high operational properties, import-independence and export-orientation. One of the promising directions in this aspect is the production of modern textile products in market conditions. Today the market needs and has the right to expect from the weaving industry a wide range of high-quality fabrics combining the optimal ratio of physical-mechanical and aesthetic-hygienic properties on the one hand, and cost indicators lying within the purchasing reach of potential consumers on the other. This is a large-scale, complex task, the solution of which requires co-ordination of efforts of scientific and engineering workers, industrialists and entrepreneurs, weaving technologists and dessinators, in a word, all those whose activity is in one way or another connected with weaving production. In turn, the efforts of scientists and scientific researchers should be directed to a considerable amount of research and design works aimed at improving the process of fabric formation on the weaving machine and modernisation of its mechanisms ensuring successful performance of this process. It is known that shuttle weaving machines of AT type are adapted for production of fabrics from yarns of small and medium linear density in weft. Due to the short length of the weft yarn in the shuttle during processing of yarns of high linear density there is a frequent change of weft packings, which leads to a decrease in fabric quality, labour productivity and equipment. Therefore, the problem of weft yarn processing on shuttle machines of AT type is very actual.

Objective of the study. The aim of this work is to produce fabrics from high line density yarns on an AT type weaving machine by modernising the weft insertion process and increasing the stock of weft packings.

Objectives of the study:

- development of a new system of weft insertion with the help of the gripper shuttle on the basis of the existing shuttle method of weft insertion on the AT-type machine;
- analytically investigate the tension of the weft and the movement of the

shuttle-gripper in the shed;

- to investigate technological and physical-mechanical properties of the new fabric structure;

- design fire hoses with water-resistant properties.

Research Methodology. In the design part, scientific and technical information on weft thread laying, the method of information analysis and thinking were used in the development of a new weft thread laying system. In the theoretical part of the thesis the methods of analytical geometry, theory of machine mechanisms, theoretical mechanics, computer technology were used in the research. Experimental researches were carried out in the weaving laboratory, where shuttle weaving machines of AT type, developed stands and devices of the certification centre at TITLP were used. Processing of experimental results was carried out by methods of mathematical statistics.

Scientific novelty of the work.

- A modernised weft insertion system based on the AT shuttle weaving machine has been developed;

- the movements of the shuttle-gripper in the shed are analytically investigated;

- The tension of weft yarns in the laying process is analysed;

- a stand and methodology for determination of thread tension for different diameters of friction guide cylinders have been developed;

- technologically and physico-mechanically investigated fabrics produced by the new method.

Practical value of the work. The practical value of the work consists in the potential possibility, based on the results of modernisation of the weft insertion process, to breathe a fresh stream in the renewal of the park of shuttle weaving machines of AT type, which will allow to expand their assortment possibilities, i.e. production of fabrics from yarns of high linear density. Besides, analytical studies of weft tension and shuttle-gripper movement in the shed, as well as the stand and methodology for determining the tension when covering its cylinders can be used in the course of weaving, theoretical mechanics and design of weaving machines.

Content of work

The introduction substantiates the relevance and relevance of the study, the purpose and objectives of the study, characterises the object and subject, shows the compliance of the study with the priority directions of development of science and technology of the republic, outlines the scientific novelty and practical results of the study, reveals the scientific and practical significance of the results obtained, implementation of the results of the study in practice, information on published works and structure of the work.

The first chapter of the paper "Literature Review and Statement of Research Problems" is aimed at analytical review of literature sources, in particular, scientific research works devoted to weaving machine feeding with duck shuttle-catch from a fixed package, mechanics of textile yarns, fabric formation process on weaving machines. In recent years, much attention has been paid to the study of the process of fabric formation and the development of the theory of the technological process of weaving. Meanwhile, in this field a lot of questions remain unsolved even now, due to a certain complexity of the technological process. All works on research of fabric formation can be divided into the following groups - works on theoretical and experimental research of the process of weft yarn surfing on the loom, works on research of weft and warp tension in the elastic threading system on the loom, works on normalisation of the technological process on the loom, works on research of fabric structure and properties in connection with conditions of its production on the loom, works on improvement of fabric formation processes and modernisation of weaving machines. The process of fabric formation is a multifactorial phenomenon. The parameters of the filling yarn surfing depend on the structure of the fabric being produced, on the operation of the tempering mechanism and warp tension, on the relative values of the stiffness coefficients of the elements in the elastic filling system, on the weaving machine filling parameters, etc. In turn, the warp tensions depend on the structure of the fabric, on the weft threading and surfing parameters, on the operation of the release and warp tension mechanism, on the relative values of the stiffness coefficients in the elastic filling system and other factors. In spite of the interconnectedness of these works, it is necessary to differentiate them in order to study more deeply the problem of fabric formation on the weaving machine. Therefore, the clarification of the essence and significance of this relationship plays a particularly important role in the study of individual issues of fabric formation. The accuracy and reality of the results obtained largely depend on the correct formulation of this question. Realising the great importance of interrelation of separate factors of the process in the formation of tissue. The question of the mechanics of material deformable yarn constrained by friction coupling is very widely covered in the literature. It is known that Leonard Euler worked extensively in this direction, where he first established the relationship between the tension of the leading and trailing parts of a flexible link sliding on the surface of a cylinder. Euler derived his formula for a cylinder mounted horizontally, with a flexible thread encircling the surface of this cylinder, positioned on it in a plane parallel to the cylinder guide. The ends of the thread hang off the cylinder and are loaded with forces T_{his} and T. The cylinder is stationary, the thread slides on its surface. The thread is weightless and non-stretchable and has perfect flexibility. Under these conditions he obtained the relation: $T = T_{0}l'$, where φ *is the* girth angle, *k is the* coefficient of friction of the flexible link on the surface of the cylinder. The paper deals with the mechanics of a weighted deformable flexible filament on a plane and other forms of guides.

A thread of length l sliding on a plane has tension from its own weight q

$$T_{pl} = q \cdot l \cdot K = q \cdot r \cdot \varphi \cdot K$$

A thread sliding on a circle, with girth arc $l = r \cdot \varphi$, has tension:

$$T_{кр} = \frac{2q \cdot r \cdot K}{1+K^2}(l^{K_\varphi} + \frac{1-K^2}{2K} \cdot \sin\varphi - \cos\varphi)$$

The above formulas do not take into account the stiffness of threads on the friction surface, as this parameter takes into account the genus and type of threads, linear plane of threads and elastic properties of threads. Therefore, it is expedient to study the yarn tension on the basis of taking into account the coefficient of yarn stiffness. The issues of weft laying are widely covered in the educational literature on weaving, which can not be said about weft laying shuttle - gripper. Weaving machines with weft laying by shuttle - grippers are produced by Textima, CBT, Adolph Saurer, Carl Zangs, etc. The main distinguishing feature of the machine system "Neumann" is weft laying on both sides of the shuttle with grippers. The filling thread is wound up from a fixed bobbin. The bobbins are installed on both sides of the machine. When the shuttle with a grip takes the thread from the left bobbin and lay it in the shed to the right, then the left thread is not cut off and its connection with the bobbin is preserved until the next threading on the left. If you stop the machine at any time, you will see that the weft threads on both sides of the machine are not cut off from the bobbins, but are directed into the shed and there worked into the fabric. The process of weft thread laying and selvedge formation is as follows (Fig.1.1). Position I - the shuttle starts moving to the left, the filling thread passes from the bobbin through the guide eye to the fabric edge, after which the shuttle can grab it. Position II - the hook has captured the weft thread and in the form of a loop introduces it into the shed. Position III - in the shed has travelled a path equal to the width of the selvedge, the knife cuts off the short end of the loop, and the long end remains in the clamp (the knife is fixed in the hook), so the short end of the loop will be earned in the selvedge. Position IV - when approaching the right side of the machine, the shuttle clamp holding the weft thread opens and the weft thread is released, the hook without weft moves into the right hook box. Position V - the shuttle is reversed and the weft is inserted from right to left. Compared to the Sulzer weaving machine, the Neumann weaving machine is considerably simpler in construction. The construction of the weft insertion mechanisms is simpler.

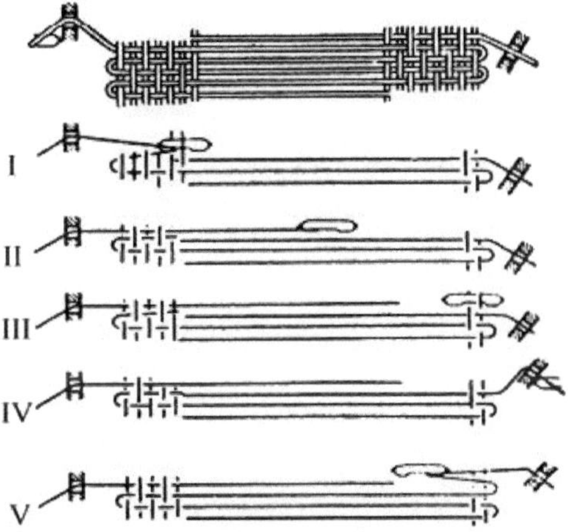

Fig.1.1 Schematic diagram of the process of weft insertion into the shed and selvedge formation with the Neumann system

The selvedges of the fabric form a normal density. The machines are used to produce shirt, lightweight suit and coat fabrics in warp and weft with a linear yarn density from 14 tex to 70 tex. In the TKDT system, the weft is laid with the same shuttle that was previously on the machine, but with the grippers attached to it. Fig.1.2 shows two variants of shuttles with grippers. The scheme of weft thread laying in the shed at the first variant of shuttle design - with grippers is presented in Fig.1.3.

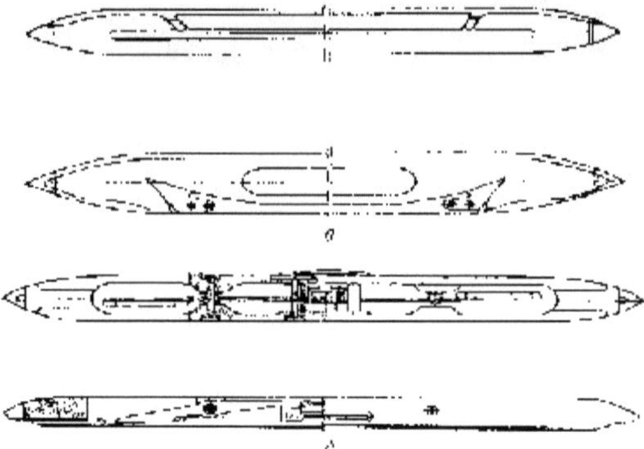

Fig.1.2 Scheme of shuttles with grippers of CKD system: a - first variant; b - second variant.

As this resulted in very large weft ends behind the fabric selvedge of up to 400 mm, a measuring device was developed. The weft yarn is laid in this way as follows (Fig. 1.3).

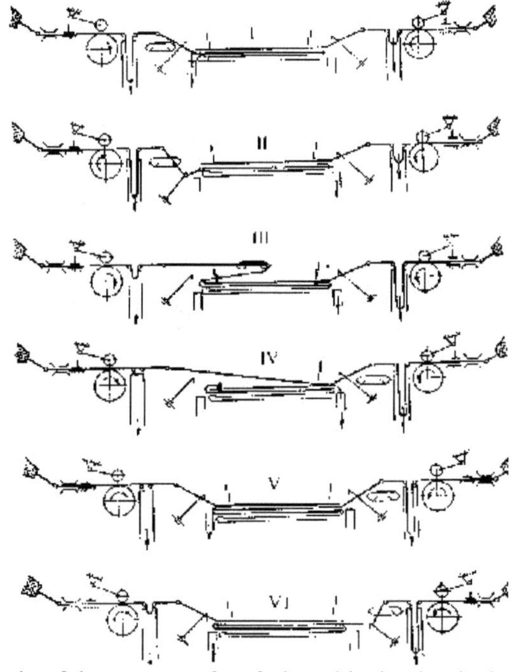

Fig.1.3: Schematic of the process of weft thread laying by the hook of the first hook
variant.

Position I - the hook is in the left hook box and the clamp is picking up the weft thread for threading in the hook. Position II - the clamp winds the weft thread into the take-up until.

Position III - the hook sews the pre-measured weft thread in the shed for one cast-off. Position IV - the scissors cut off the weft thread laid 30 mm deep into the shed. The grommet ends from left to right. Position V - the clamp picks up the weft thread for threading in the hook grip when laying from right to left. The measuring device measures the weft thread for one threading. Position VI - the clamp brings the weft thread into the hook grip, and all is repeated on the right side, etc. Fig. 1.4 shows the scheme of weft laying in the second version of the design of the shuttle with grippers. The filling thread is wound from the bobbin 1, passes the thread tensioner, working from the cam 2, then the clamp, which works from the cam 3, thread guide 7 and falls into the nozzle 6. Switches 4 supply and close the air in nozzle 6.

The bobbin 5 picks up the weft thread in flight and passes it through the shed. The scissors 8 cut off the weft thread after each shedding. Edge formation is not solved in this scheme.

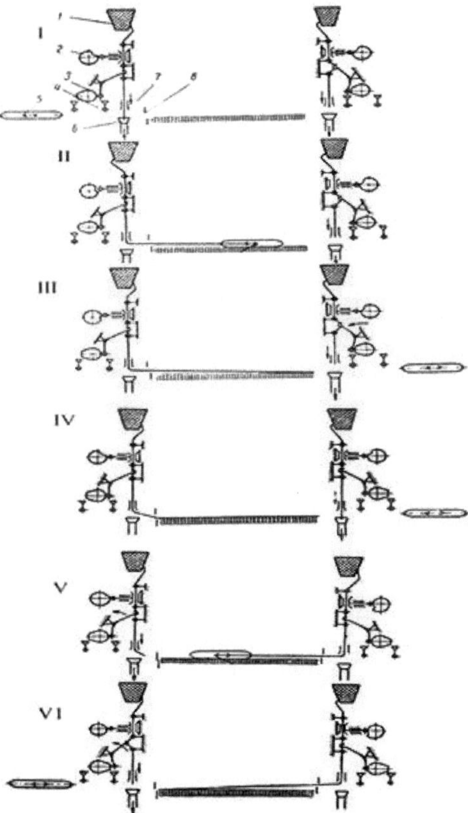

Fig.1.4: Schematic of the process of weft insertion by the second hook

variant.

Using the weft insertion of the Adolph Saurer weaving machine with the Adolph Saurer pick-up system, the machine is designed for weaving fabrics from yarns made from natural and man-made fibres as well as from a mixture of different fibres. The loom can also be used to process combed cotton yarns. The company is developing a machine model for the production of filament yarns and low number yarns. The machine operates at a speed of 300 weft passes per minute with a filling width of 110 cm, the quality of edges on both sides of the fabric is satisfactory. Weft is laid by shuttle with grippers from conical or cylindrical bobbins (Fig.1.5), located at the edges of the fabric. First, the weft thread is laid from one bobbin (for example, from the right, see position I, Fig. 1.5), and the length of thread for one throw is measured in advance. Then the weft thread is laid from the left bobbin (position II), after which it is laid again from the right bobbin without being cut off and forms a loop with the previously laid weft thread (position III). The thread is laid in the same way on the left side (position IV), etc.

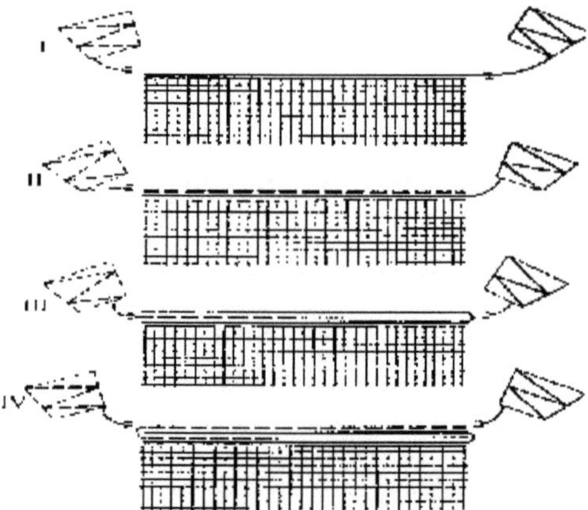

Fig. 1.5: Diagram of the weft insertion process on the Adolph Saurer weaving machine.

By inserting the weft yarn into the shed in the form of a loop alternately on the right and left, a strong selvedge can be created. As a result of the use of measuring devices, the length of the weft ends protruding beyond the fabric selvedge is no more than 3-4 mm. With the use of weft insertion by the Carl Zangs shuttle-gripper system, the weaving machine is designed for the production of natural silk, synthetic fibres, taffeta, lining fabrics, umbrella fabrics, as follows (Fig.1.6). Position I - foot 1 opens the hook gripper 2. The suction nozzle 3 is lowered and introduces the fine thread of the hook. The threading system 5 promotes tensioning of the weft thread and its correct position in the hook catcher. When trolling, the hook passes the thread behind it, which is rewound from the bobbin 6. Position II - the shuttle box on the other side of the machine is at some distance from the edge. Between the shuttle box and the selvedge is a lever for the reserve thread 7, which measures the appropriate length of weft for the subsequent threading. This reserve weft in the form of a loop is above the flight path of the shuttle. The scissors 8 cut the weft thread, the reins are raised and the reed nails the weft. Position III - device 9 holds the end of the weft thread against the suction nozzle 3, which takes the initial position. Position IV - device 10 removes the loop from lever 7, and the hook is thrown through the shed. Position V - lever 7 is lowered again.

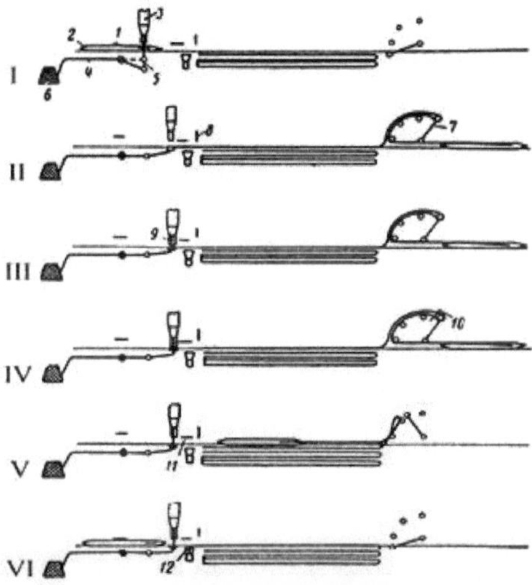

Fig.1.6. Schematic of the process of weft thread laying by shuttle
with Carl Zangs
type system gripper.

Foot 11 opens the hook catcher, releasing the end of the weft thread. Position VI - this end is held by nozzle 12 during the lifting of the bobbins and surfing with the reed. Scissors 8 cut off the protruding end, which is immediately sucked off. The fabric produced on this loom has a regular selvedge on one side and a leno selvedge on the other side. The mechanisms of the Zangs loom are simple and can be mounted on a standard shuttle loom frame. The depth of the shed is retained. The fighting mechanism, automatic warp tension regulator, warp monitor, etc. remain unchanged. The new mechanism is installed on the shuttle loom in the form of separate units. A weaving machine with a shuttle with a gripper can be supplied with a conventional shedding mechanism, i.e. with an eccentric mechanism, a remise-lifting carriage, a jacquard machine. The machine can produce fabrics from yarns with a thickness of 100-3.3 tex (No. 10-

300), from the lightest to fabrics weighing 400 g/m2. The weft reel, the reserve reel and the leno selvedge unit are mounted on the frame of a conventional weaving machine. At high speed, the reeling from the stationary bobbin on weaving machines with gripper shuttle is uneven (jerky) and therefore creates a strong tension causing thread breaks. The shape of the bobbin plays a major role in winding. Sometimes synthetic yarns are wound in a wet or damp state, the layers of yarn are tightly adhered to each other, and therefore different tension occurs when winding. To improve the winding of the yarn, the cone bobbin is mounted on a spindle rotating in the opposite direction to the winding of the yarn. The spindle is driven from the centre shaft via a bevel gear. The number of revolutions of the spindle should be such that the length of the winding thread for 1 cycle was less than the width of the threading. When using the Clayes-type shuttle-capture system, it is possible to produce cotton fabrics from yarns of medium linear densities. On machines with shuttles with gripper shuttles of Clayes type (Fig.1.7) the shuttle with gripper has the same shape as on the machine of Neumann system and works in the same way, however, the process of selvedge formation is different. On the Neumann machine, the shuttle is only used for edging with the help of the cutter bars, which do not work simultaneously with the main cutter bars. When leaving the shed, the shuttle cuts and lowers the sinker just opposite the bending end of the preceding sinker. On Trullas - Clayes machines, the hook cuts the weft thread tautened by the suction device a little later when leaving the shed, so that the end of the weft thread leaves the shed well taut and doubles the bending end of the preceding weft.

Fig.1.7. Conventional shuttle and shuttle-capture of Clayes system.
This results in a slight thickening, as the twin end of the weft is bent at the selvedge. However, this thickening does not interfere with further processing of the fabric, as it is considerably smaller than on Sulzer machines. The shuttle, entering the shed, puts the bending end of the weft thread in question in the same way as it happens on the machines of the system "Neumann". There is also work on modernisation of shuttle machines. The firm G. Kapps (France) is modernising old machines for the production of very inexpensive fabrics and blankets with low weft yarn numbers for weft feeding from fixed bobbins, as the purchase of new expensive machines is not feasible in this case. In order to

reduce weaving costs, such products should be produced on weaving machines without bobbins. Automatic looms with cob change are not well suited for this branch of weaving. For the same reason it is not possible to increase the speed of the looms. Automatic shuttle looms are also not a solution, because with 835 tex weft yarns (No. 1, 2) and a filling width of more than 3 m, the cob is triggered in just over a minute. The yarn should be sufficiently taut and have a breaking strength of 400 to 600 G. Shuttle mechanisms with a grip, spread in recent years in France, requires the use of strong shuttles. The new mechanism, which the company Kapps used on the old machines, carries out the laying of weft thread with a tension of 350G. This mechanism consists of three main elements: a shuttle with a grip, which has the shape of a conventional shuttle, and a mechanism for selecting the colour of the weft thread for the production of three-colour plaid fabrics. Weft yarns of 111 tex (No. 9) can be processed on a machine equipped with this mechanism, but particularly good results have been obtained with cotton yarns of 16.7 tex **x** 2 (No. 60/2). The mechanism works as follows. During the stop of the shuttle with a grip in the shuttle box, the needle introduces the end of the weft yarn from the side into the shuttle through a hole made in the box. The needle puts the thread inside the shuttle with a grip, which has the same dimensions as the old shuttle with a bobbin, driven by the usual combat mechanism, so that the shuttle with a grip can not cause excessively high tension weft thread when unwinding it from the bobbin, it is necessary to use a special unwinding mechanism, instantly issuing the full length of weft thread required for one sweep. Working tests of the machine must be carried out for each new type of shuttle and for each machine with a large filling width. After the first tests have been carried out, the working samples of the machines will have to be used in the various branches of weaving. It is very important for practical purposes to know whether the gripper shuttle system can be used with high yarn counts and on machines with large filling widths. In summary, the classification of weft insertion methods into the shed has been clarified by the introduction of weft insertion by a hooked shuttle. The existing systems of weft insertion by shuttle-capture are complicated in design and require a number of additional mechanisms for fabric formation. It is reasonable to develop a new system of weft insertion by shuttle-gripper on the basis of shuttle weaving machine of AT type.

In the second chapter a new system of weft insertion by the shuttle-gripper is developed. Modernisation of old machines of AT type for production of inexpensive fabrics (technical) and blankets, where weft yarn of high linear density is used, with weft feeding from fixed packings is reasonable, as weaving production costs for these products are reduced, and frequent change of bobbins or shuttle with bobbin affects fabric quality and productivity of labour and equipment. Therefore, below we will consider mechanisms for weft insertion from fixed bundles with the help of shuttle hooks. Modernisation of weft insertion was carried out with maximum use and minimum change of mechanisms. The weft yarn is fed from two fixed bobbins installed on the right and left sides of the machine, Fig.2.1 shows the feeding of weft yarn from the bobbin located on the left side of the machine. Fine thread from each bobbin is fed to the shuttle gripper sequentially from the right and then from the left side by means of two (right and left) mechanisms for feeding the bobbin. The mechanism of feeding utochina is an eye (Fig. 2.2), which is mounted on the bar remizki 1. When lowering the remizki down, the eye 2 with a threading 3 threaded in it will set the thread on the line of movement of the shuttle-capture 5, which will catch the threading and put it in the shed. In the reverse movement of the shuttle, another remizka will set the weft on the line of action of the gripper and the weft will be laid through the shed. Consequently, the setting (lowering of remise) of the weft on the line of capture must be from the side of fight (acceleration) of the shuttle-capture. Since the combat cycle is equal to two, the delivery cycle of the sinker is also equal to two. The most rational system of feeding the weft in plain weave, as the shedding cycle is equal to two.

Fig. 2.1: Feeding the weft yarn from the fixed bobbin.

Fig.2.2: Thickness feeding mechanism

When the hook exits the shed, the remise enters the scoring phase, and the eye with the sinker moves upwards, which causes the sinker to leave the hook grip. The ends of the wefts remaining at the fabric edges are 60-100mm long. In conclusion, it can be noted that the mechanism of feeding the weft is very simple

by design and has an original technical solution, as it replaces the whole system (cams, levers, rods, etc.) of parts. The mechanism of weft yarn withdrawal from the hook grip (Fig.2.3) is based on the fact that when lowering the bobbin threader down from the side of the bobbin, the weft yarn can freely wind off the bobbin.

Fig.2.3. Mechanism of weft thread withdrawal from the hook grip.

At the end of weft laying, the weft threader moves upwards (changing its position according to the weave pattern) and starts to clamp the weft thread with its elastic plate, and there is a gradual clamping of the thread and, as a consequence, a gradual increase in the tension of the weft thread. As a result, utochina slips off the grip of the hook at the opposite edge of the fabric. The beginning of the action of the flat spring on the weft and the amount of clamping of thread, regulate the movement of the flat spring on the remizki vertical at the eye of the weft thread feed. Consequently, feeding and braking of the weft thread with the help of the eye and flat spring using the movement of the weft thread guide (downwards on the side of the weft and upwards at the end of weft laying), determines the simplicity of the construction and reliability of the mechanism. The shuttle of the modernised weaving machine with fixed weft insertion differs from the shuttles of AT type weaving machines. Structurally, the shuttle is a wooden beech bar 1 (Fig.2.4), on the ends of which steel toes 2 are fixed, where there are grippers, which serve for gripping the weft thread 3 and the weft thread 3.

pulling it through the pharynx.

Fig.2.4 Modernised Shuttle Gripper

The middle part of the shuttle is laid out in the form of an oval to better guide the weft thread to the grippers. The angle between the back wall and the lower plane of the shuttle corresponds to the angle between the reed and the slip of the batana and is 90^0 . The ends of the wooden bar and the toe are cut at an angle corresponding to the position of the weft on the section from the down of the fabric to the eye of the weft feed mechanism, which ensures the reliability of the weft thread capture by the hook. It is also possible to position the gripper in the form of a vertically mounted rod on the toe - shuttle. By design, the rod can be made of rigid material or elastic material. Also on the rod can be mounted easily rotatable sleeve 1, in contact with the utochina to reduce friction between the utochina and the gripper guide. Below are the parameters of the shuttle-gripper: total length with toe - 440mm;

Back wall height - 36mm; Front wall height - 30mm; Hook width - 44mm; Grip height - 5mm; Angle between reed and slider - 900; Weight of hook - 400g. The mechanisms of the weaving machine can only work correctly if they are co-ordinated. The operation of the mechanisms is coordinated by means of a cycle diagram, according to which the machine mechanisms are adjusted. The cycle diagram (Fig. 2.5) gives an idea of the effect of the mechanisms for two revolutions of the main shaft of the machine.

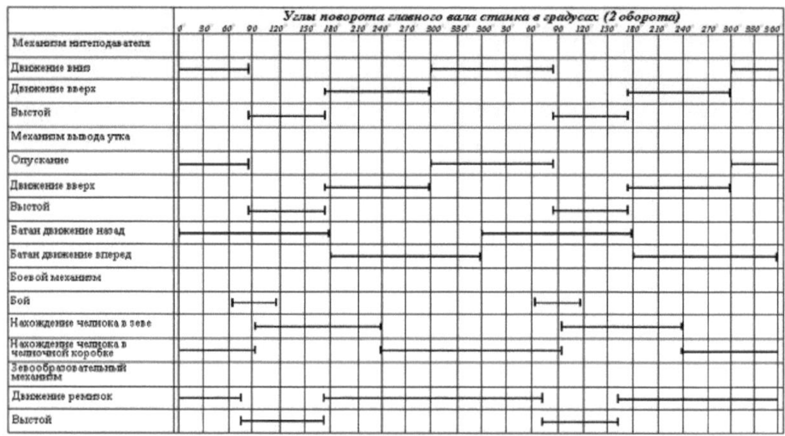

Fig.2.5 Cycle diagram of mechanisms operation.

The cycle diagram shows the stages of movement and operation of the mechanism for feeding the weft thread to the hook, the mechanism for withdrawing the weft from the hook, the combat mechanism, the batana and the shedding mechanism. The work of the first and second mechanisms depends on the diagram of movement of remiz, as the elements of these mechanisms are installed on the remiznaya frame. The operation of the batane mechanism can be broken down into two stages. The first stage is the backward movement of the batane, which corresponds to 0^0 to 180^0 , and the forward movement of the batane, which corresponds to 180^0 to 360^0 rotation of the main shaft of the machine. The operation of the combat mechanism is carried out in three stages. The first stage is the beginning and the end of the combat, which corresponds to 70^0 to 100^0 rotation of the main shaft of the machine. The second stage - movement of the shuttle-catch in the shed with weft thread, which corresponds to 100^0 to 240^0 rotation of the main shaft of the machine. The third step is to locate the hook-gripper in the shuttle box from 250^0 to 100^0 of the next revolution of the main shaft of the loom. The shedding mechanism can also be broken down into three stages. The first stage - the movement of the doffer to open the shed corresponds from 300^0 to 180^0 of the next revolution of the main shaft of the loom. The second stage - movement of the shearer from 80^0 to 170^0 turn of the main shaft of the machine tool. The third stage is the movement of the slitter to close the shed, which corresponds from 170^0 to 300^0 of the main shaft rotation of the machine tool. The work of the mechanism for feeding the utochina to the hook grip can be divided into three stages. According to the diagram, the work of this mechanism corresponds to the work of the shedding mechanism. The first stage - feeding weft thread, lowering the eye, on the line of capture hook, which corresponds to 300^0 to 80^0 the next turn of the main shaft

18

of the machine. And feeding weft thread comes from the remizki, lowering down exactly from the side of the battle. The second stage - the eye in the lower position from 80^0 to 170^0 rotation of the main shaft of the machine. The third step is to move the eye upwards from 170^0 to 300^0 rotation of the main shaft of the machine. The cycle is then repeated for another eye on a different threader. The work of the mechanism of withdrawal of weft thread from the shuttle can be divided into three stages. The first stage - downward movement together with the remizka from 300^0 to 80^0 , which determines the free winding of the weft from the bobbin, and the moment of capture weft thread shuttle. The second stage - stoop in the lower position from 80^0 to 170^0 , where the movement of the shuttle-gripper in the shed is laying the weft and its free winding off the bobbin. The third stage - upward movement with remizka from 170^0 to 300^0 , which causes an increase in the tension of the weft thread, ie the braking of the weft and slipping it off the grip of the shuttle. Other mechanisms of the shuttle machine AT work without changing the cycle diagram, such as the mechanism of the weft fork, automatic bobbin changer, the mechanism of release and warp tension, the mechanism of withdrawal and winding of the fabric, etc. To sum up, it should be noted that the shuttle weaving machine has been modernised by using the nature of movement of the reims by installing on the reims the mechanisms for feeding and withdrawal of the weft to the shuttle gripper. the parameters of the shuttle gripper have been determined. the cycle diagram of interaction of the main mechanisms of the modernised machine has been developed.

In the third chapter, the studies of the weft tension and movement of the shuttle-gripper in the shed are carried out. Having received the maximum speed during the period of acceleration in the box, the shuttle continues its way in the shed at a constant decrease in speed, due to the forces of resistance, slowing down its movement (Fig.3.1). To the forces of resistance should include friction of the shuttle-capture on the warp and slip, friction on the reed, air resistance, tension of the weft thread. Neglecting the air resistance and tension of the weft thread, let's consider the influence of the main forces acting on the shuttle-gripper during its movement. Pressure on the shuttle-gripper from the reed side occurs under the influence of the movement of the batana. Since the movement of the shuttle-gripper occurs during the period of the main shaft from 90^0 to 270^0 , it can be considered that in this part of the main shaft revolution is pressing the shuttle to the reed. The action of inertial forces, pressing the shuttle to the reed for the corresponding part of the crankshaft revolution, is essential in that the shuttle-gripper makes its movement within this part of the revolution and receives due to this force closure, guiding it along the reed. On the other hand, the pressure of the hook-gripper on the reed creates friction between the rear wall of the hook-gripper and causes wear of the hook-gripper and the reed.

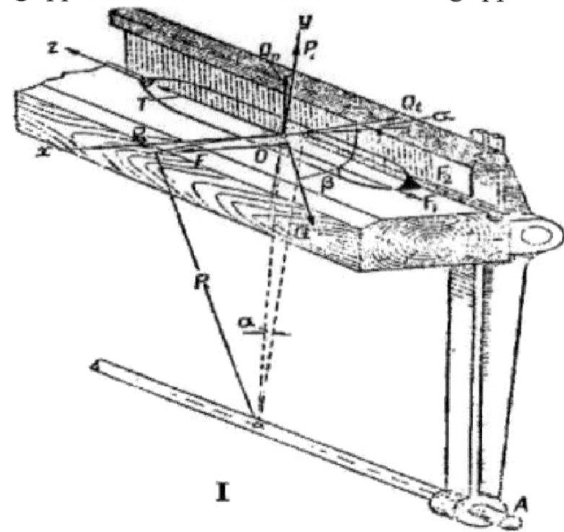

Figure 3. 1. Forces acting on the shuttle-gripper during its movement along the slick.

In addition to tangential acceleration, the shuttle-catcher has normal acceleration in forward motion, hence the normal component of the inertia force, or centrifugal force. Under the influence of these forces and weight of the shuttle-gripper there are reaction forces from the side of the slip and reed, causing friction forces of the shuttle-gripper against the slip and reed. Let's introduce the following notations: **G** - weight of the shuttle-gripper; Q_n - normal component of inertia force of the shuttle-gripper in its transfer motion, i.e. centrifugal force; **Qt** - normal component of inertia force of the shuttle-gripper in its transfer motion, i.e. centrifugal force; Q_t - tangential component of the force of inertia in the portable movement; F_1 - friction forces of the shuttle-gripper on the threads of the warp in its portable movement; **F2** - friction forces of the shuttle-gripper on the reed in the relative movement of the shuttle-gripper; **f1** - **friction** coefficient of the shuttle-gripper against the warp thread; **f2** - **friction** coefficient of the shuttle-gripper against the reed; **ω** - angular velocity of the batana; **a** - acceleration of the shuttle-gripper; a_p - normal acceleration in the transfer motion of the shuttle-gripper; a_t - tangential acceleration in the portable motion of the shuttle-gripper; a_r - acceleration of the relative motion of the shuttle-gripper; a_k - Coriolis acceleration; P_1 - reaction force of the slip of the bataan, perpendicular to the plane of the slip; **P2** - reaction force of the reed, perpendicular to the plane of the reed; **T** - inertia force in the relative motion of the shuttle on the slip of the bataan; **m** - mass of the shuttle; **R** - distance from the centre of gravity of the

20

shuttle to the axis of rotation of the bataan. The forces applied to the shuttle-capture are shown in Fig. 3.1, and the components of the inertia force of the transfer motion of the shuttle Q_t and Q_n are applied in the centre of gravity. Assuming that the shuttle-capture moves along the slope in the direction parallel to the axis of rotation of the batana, we obtain in this case the Coriolis acceleration equal to zero, i.e. $a_k = 2 \cdot [\vec{\omega} \cdot \vec{v}_r] = 0,$ (3.1), *since the* vector product $[\omega\text{-}_{vr}]$ is equal to

zero due to parallelism of vectors $\vec{\omega}$ and $\vec{v}_r$. Therefore Coriolis force of inertia in the considered case is absent. Let's bring the considered system of forces to

the centre of gravity of the shuttle. Obviously, the main vector $\vec{R}$ of the system, according to the Dalembert principle, will be equal to zero, i.e.

$$\vec{R} = \vec{G} + \vec{P_1} + \vec{P_2} + \vec{F} + \vec{F_1} + \vec{F_2} + \vec{Q_n} + \vec{Q_t} + \vec{T} = 0 \qquad (3.2)$$

In this equation: $G = mg$, $F = f1\ P_1$, $\quad F2 = f_2\ P_2$, $Q_n = m\omega 2R$,

$$Q_t = mR\ d\omega\ /\ dt, \quad T = mor, \quad F_1 = f1\ P_1.$$

We introduce the coordinate system ox, oy, oz; the direction of the ox axis
is perpendicular to the
reed plane, oy - perpendicular to the slab plane, and oz - in the direction of the relative velocity of the shuttle. This coordinate system is rectangular under the assumption that the reed and slalom planes are mutually perpendicular. Having projected equality (3.2) on axes, we obtain:

$R_x = P2 + F + Q_t\cos \alpha + Q_n\sin \alpha - G\cos \beta = 0$ (3.3)
$R_y = P1 + Qn\cos \alpha - G\sin \beta + Qt\sin \alpha = 0$ (3.4).
$R_z = - Fi - F2 + T = 0$ (3.5)
It follows from equations (3.3) and (3.4):
$P1 = G\sin\beta - Qn\cos \alpha - Qt\sin \alpha$ (3.6)
$P_2 = G\cos \beta + Qt\cos \alpha - Qt\sin \alpha - F$ (3.7)
$fi\ P_i = fi\ (G\sin \beta - Qn\cos \alpha - Qt\sin \alpha)$ (3.8)
Hence: $P2 = G\cos \beta + Qt\cos \alpha - Qt\sin \alpha - fi(G\sin \beta - Qn\cos \alpha - Qt\sin \alpha) = =$
$= G\ (\cos \beta - fi\sin \beta) - (Qn - fi\ Qt)\sin \alpha + (Qt + fi\ Qn)\cos \alpha;$
then: $F2 = f_2\ P2 = f_2\ [G\ (\cos\beta - fi\sin\beta) - (Qn - fi\ Qt)\sin d + Q + f\ Qn)\cos o.]$ (3.9)
During the period of movement of the shuttle in the shed the bateau moves with insignificant angular velocity about its rear position. Therefore, the magnitude of the centrifugal force of the shuttle $Qn = \omega 2R$ can be neglected. In addition, the angle α is very small, and we can take $\cos\alpha = 1$, sina. On the basis of this we will have:

$$P_i = G\sin \beta \quad (3.10)$$

$$P2 = G\ (\cos \beta - fi\cos \alpha) + Qt \quad (3.11)$$

It follows from equations (3.10) and (3.11) that the magnitude of the forces P_1

and $P2$ depends on the angle β, which varies with the crankshaft rotation angle φ, measured from the front dead centre position of the crank. Therefore, it would be possible to express $P1$ and $P2$ analytically as functions of the angle φ of rotation of the main shaft. We can consider that the magnitude of the force $P1$ is approximately equal to the weight of the shuttle, and the force **P2** is equal to m^at for the period of movement of the shuttle in the shed. Having in mind the equality (3.5), the basic equation of dynamics for the movement of the shuttle on the slip can be written in the form: im12 z / d t₂ = - F1 - F2 (3.12).

Substituting in the last equation **d2z/dt2** by **dV/dt** and multiplying the left part by **dz/dz** we obtain:

$$m\frac{dV_2}{dt}\cdot\frac{dz}{dz}=-F_1-F_2;$$

$$mVdV = - (F_1+F_2)\ dz,$$

By integrating both parts of the equation, we obtain:

$$m\frac{V_2-V_1}{2}=\int_{z_1}^{z_2}(F_1+F_2)dz.$$

Considering the forces $F1=f1\ P1,\ F2=f2\ P2$
Constant in magnitude, we obtain:

$$m\frac{V_2-V_1}{2}=-(F_1+F_2)(z_2-z_1)=-(F_1+F_2)S,$$

where S is the path travelled by the shuttle in the shed.

$$2[-(F_1+F_2)S]=(V_2-V_1)m$$

$$\frac{2[-(F_1+F_2)S]}{m}=V_2-V_1$$

$$V_2=V_1-2S(F_1+F_2)/m \qquad\qquad (3.13)$$

Having determined by the above method forces $P1$ and $P2$, and therefore, and the sum of forces $F1 + F2$, using the equation (3.13), the speed of the shuttle in any place of its movement in the shed. Table 3.1 shows the calculation of the speed of the shuttle-capture in the shed at fi=f-=0.2; t=0.041kg; P1=0.4kg; at=5m/sec² at V^^M/sec; at=3m/sec² at V_1 =10 m/sec; at=1m/sec² at V_1 =5 m/sec, F_1 =f₁ P_1 ; F =f₂₂ P_2 =f₂ mat

Table 3.1.

№	Shuttle-gripper speed	Shuttle-gripper speed in the shed, m/s				
		The path travelled by the shuttle-catcher in the yawn, м				
	V1, m./sec	0,2	0,4	0,6	0,8	0,10
1	15	14,4	13,8	13,2	12,6	12,0
2	10	9,0	7,9	6,9	5,9	4,9

3	5	4,2	3,3	2,5	1,6	0,7

Fig. 3.2 shows the graphs of change of the shuttle-gripper speed in the shed as a function of the path travelled by it along the width of the machine dressing.

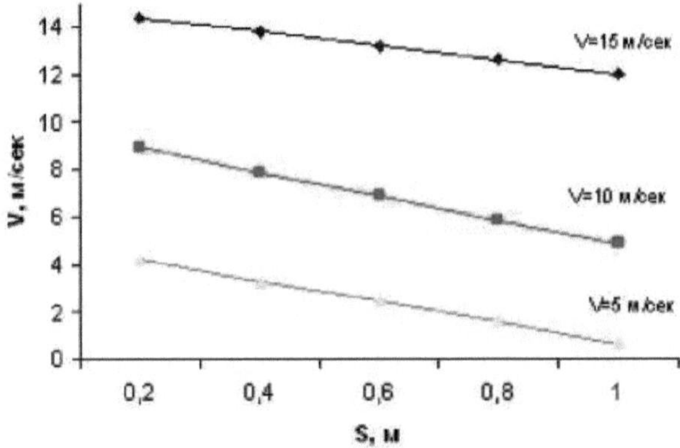

Fig.3.2. Graphs of change of speed of the shuttle-gripper in the shed depending on the path travelled by it along the width of the machine dressing.

The *friction* force $f_1P_1 = F_1$ creates a torque with respect to the centre of gravity of the hook equal to M1=F1b (2b - height of the hook), under the influence of which the uniformity of pressure distribution on the hook from the side of the batana squeeze is disturbed: the pressure on the front part of the hook increases and on the back part decreases. Therefore, the underside of the hook is more worn at the ends than in the middle. Fig.3.3 shows an approximate distribution of pressure on the lower plane of the hook along the curve mn instead of the assumed uniform distribution. The equidistance of this pressure, obviously, will pass to the left of the centre of gravity of the shuttle. The moment of the equidistant P_1 will balance the moment M1=F1 b, ie.

Pipi = F_1 b or: *Pipi* = *fi* P_i b. Where **pi-** *deviation* **of** force P_i from the centre of gravity **of** the shuttle. Reducing by P_1, we get: $p1 = f1\,b$ (3.14)

The last equality shows that the value of p_i deviation of the equal force from the centre of gravity of the hook, and hence the uniform adjoining of the hook to the slip, depends on the height of the hook: the smaller P, the more uniform will be distributed the pressure of the hook on the warp threads.

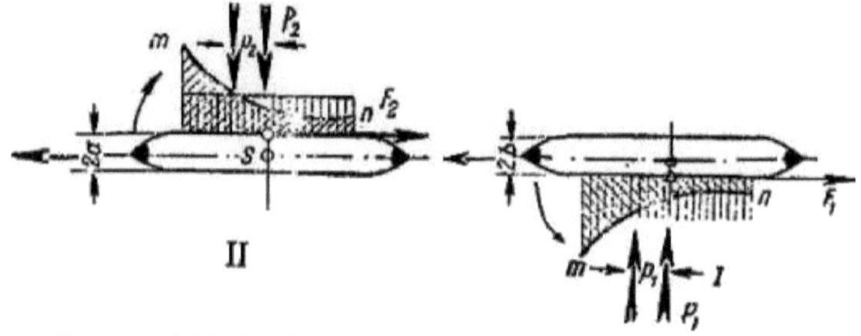

Fig.3.3 Influence of friction forces on shuttle movement

From the action of friction force F2 there is a moment F2 (2a - width of the hook), under the influence of which the uniformity of pressure distribution from the reed side is disturbed; the pressure in the rear part of the hook decreases, in the front part - increases (Fig.3.3). As well as in the previous case, the equidistance of pressure on the hook from the reed shifts from the centre of gravity to the left in the direction of movement of the hook. From the equilibrium condition follows: *P2 p2=F2 a, P2P2 = f2Pa, $p2 = f2a$,* (3.15). (**whereP2** - deviation of force *P2* from the centre of gravity of the shuttle), i.e. deviation *p2* from the centre of gravity of the shuttle is determined by the coefficient of friction *f2* and the width of the shuttle 2a: the wider the shuttle, the more uneven will be distributed its pressure on the reed, the more will be triggered ends of the shuttle and the stronger will be the reed wear. From the equations (3.14) and (3.15) we can conclude: the smaller the height and width of the shuttle-gripper, the more correct and stable will be its movement. In the above studies the influence of the weft tension on the shuttle-gripper flight was not taken into account, because in most cases of loom operation this influence is not great. However, it can be quite important when low weft yarn numbers are used. In these cases, it is not uncommon to see the shuttle flying out of the shed under the influence of the tension of the weft yarn and the instantaneous fixation of the yarn in the moving shuttle-gripper. In addition, there may be a slippage of thread from the grippers and its loss in the shed. Fig. 3.4 shows three possible directions of the weft thread relative to the centre of gravity of the hook. Let's assume that the weft thread tension is directed along the thread. In position I, the tension of the thread gives a moment that rotates the hook in the direction of the arrow and increases the pressure on the reed of the front end of the hook. The least pressure of the front end of the hook against the reed is transferred to position III.

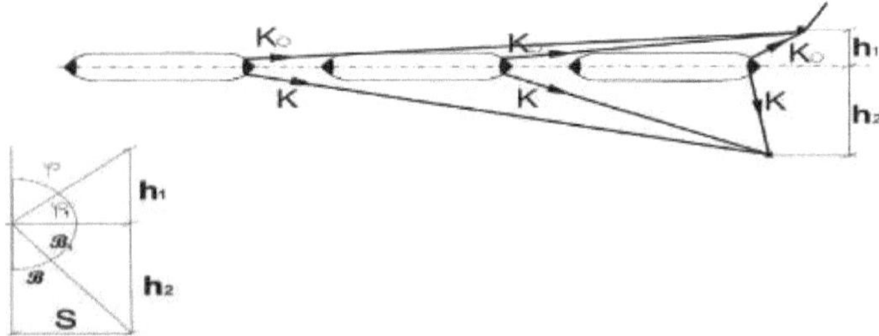

Fig.3.4 Schematic of movement of the hook-gripper in the shed: where S - the magnitude of movement of the hook-gripper; K0 - the incoming force of thread tension; K - the escaping force of thread tension; h1 - the distance from the reed to the toe of the hook; h2 - the distance from the toe to the fabric edge.

According to L. Euler, the ratio between the incoming (κo) and escaping (K) branches has the following expression and depends on the friction angle (α) and the coefficient of friction of the thread on the guides of the shuttle-catcher.

$$Ko= Ko\ exp(fa) \quad (3.16)$$

According to Fig.3.4, the friction angle α is defined by $\alpha = \varphi + \beta$ (3.17) whence $\varphi=(90^0 - \varphi 1)$ (3.18), $\beta=(900- \beta 1)$ (3.19), $tg\ \varphi 1= h1/S$ (3.20), $tg\beta i=h2/S$ (3.21). Taking into account that hi, h2 values are constant and can be determined practically on the filling machine Sh=22mm, h2=128mm, and the value S displacement of the shuttle-gripper in the shed, it is possible to determine the values of angles φ, φι, β, βι and ultimately the friction angle α of the sink about the gripper shuttle. Table 3.2. shows the results of calculations of angles φ, φι, β, βι and the angle of friction of the thread α on the guides depending on the position of the hook-gripper in the shed.

Table 3.2.

Results of calculations of angles φ, φι, β, βι and yarn friction angle α against the guides in the

depending on the position of the shuttle-gripper in the shed.

№	Angles, in degrees	Shuttle movement in the shed, m.						
		0	0,1	0,2	0,4	0,6	0,8	1,0
1	φ1	90	12	6	3	2	2	1
2	φ	0	78	84	87	88	88	89
3	β1	90	52	33	18	12	9	7
4	β	0	38	57	72	78	81	83
5	α	0	116	141	159	166	169	172

The analysis of Table 3.2 shows that the maximum angle of friction of the weft against the hook grip takes place at the exit of the hook from the shed. Euler's formula (3.16) gives the same thread tension K for a given girth angle α and the tension of the incoming branch **K0** regardless of the shape of the guide grip on which the thread is located. For example, for large cylinders with different diameters and at the same girth angles, the tension **K** is the same. Undoubtedly, the thread tension cannot be the same for different shapes of the cylinder guide along which the thread is arranged. It may be greater for some cylinder shapes and less for others. Let us consider variants when the cylinder guides are straight line, circle. Taking into account some changes we have made in the formulae we have. For a thread sliding along a plane, we have

$$K = Ko + Knlf \qquad (3.22)$$

where: Ko - tension of the incoming thread weft; Kn - stiffness of the weft yarn, depending on the type of fibre and linear density of yarn, cN/mm; l - length of thread, sliding on the guide of the hook gripper mm; f - coefficient of friction of the thread on the guide cylinder of the hook cylinder.

A thread of length l, equal to the product of the friction radius r by the friction angle α, sliding on a plane has a tension Kzp,

$$Kzp = Kn + f = Kn \cdot r \cdot l - \alpha f \qquad (3.23)$$

Consequently, the tension of the sinker in the escaping branch when sliding on the plane has the form, $K = Co + Kzp \qquad (3.24)$

A thread of length l, sliding along the circle at the arc of coverage equal to r- l has tension

$$K_{30} = \frac{2 \cdot K_H \cdot r \cdot f}{1 + f^2} \left(exp\, f \cdot \alpha + \frac{1 - f^2}{2 \cdot f} \cdot sin\, \alpha - cos\, \alpha \right) \qquad (3.25)$$

The tension of the sink in the escaping branch as it slides along the circumference has the following form $K = Co + Kzo \qquad (3.26)$

At $\alpha = \pi$ the additional tension of the thread caused by the action of the shuttle-gripper is $Kzp = Kn \cdot g\, \pi - f \qquad (3.27)$

$$K_{30} = \frac{2 \cdot K_H \cdot r \cdot f}{1 + f^2} \left(exp\, f \cdot \alpha + 1 \right) \qquad (3.28)$$

Below we plot (Fig. 3.5) the graph of change of additional tension of the sinker as a function of the friction angle. In the calculation of equations (3.27) and (3.28) $\pi\pi\pi\pi\pi\ \pi$

Accepted: $f = 0,2;\ \alpha = \pi,\ \dfrac{\pi}{2},\ \dfrac{\pi}{4},\ \dfrac{\pi}{6},\ \dfrac{\pi}{8},\ \dfrac{\pi}{10}$.

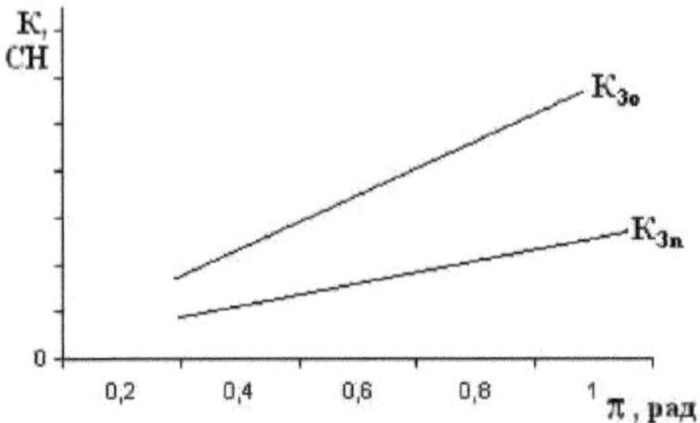

Fig. 3.5. Graph of change of additional tension of the sinker as a function of friction angle.

It follows from Fig. 3.5 that, depending on the type of guide (plane or cylinder) through which the thread is thrown, we obtain different values of thread tension at the same value of thread stiffness, friction radius, friction coefficient and initial tension (on the running branch of the thread). For a thread sliding on a plane the tension of the sinker is less than for a thread sliding on a circle. Table 3.3 - 3.5 shows the results of calculations of sinker tension on the shuttle grip and running branch when the yarn moves on a plane, on a fixed cylinder and on a rotating cylinder depending on the friction radius, friction angle and friction coefficient for yarn,

having stiffness 10sn/mm (linear density of yarn 320 tex). The values of friction coefficient were taken according to experimental studies. The analysis of tables 3.3-3.5 shows that it is advisable to use as a gripper weft thread on the shuttle rotating cylinder. Since it is impossible to realise the movement of the weft thread on the plane as a gripper of the weft on the shuttle, and the movement of the thread on the fixed cylinder creates a large braking effect on the weft, which can cause premature slipping of the weft from the gripper or change the trajectory of the shuttle.

Table 3.3.

Influence of the angle and friction coefficient on the tension of
the thread when the
thread moves
along the plane. When the friction radius r=1.

№	Tension Utochiny, CH	Coefficient of friction of the thread on the hook grip, f	Angle of friction of the sink against the gripper,^{a} , deg.					
			116	141	159	166	169	172
1	On the shuttle's	0,19	3,5	4,2	4,7	5,0	5,1	5,2
	grip	0,28	5,1	6,2	7,0	7,4	7,5	7,6
2	In the	0,19	8,5	9,2	9,7	10,0	10,1	10,2
	descending	0,28	10,1	11,2	12,0	12,4	12,5	12,6

№	Tension Utochiny, CH	f	116	141	159	166	169	172
	branch of the thread							
At friction radius r=2								
1	On the shuttle's grip	0,19	6,9	8,4	9,5	10,0	10,1	10,3
		0,28	10,2	12,4	14,0	14,7	14,9	15,1
2	In the descending branch of the thread	0,19	11,9	12,4	14,5	15,0	15,1	15,3
		0,28	15,2	17,4	19,0	19,7	19,9	20,1
With friction radius r=3								
1	On the shuttle's grip	0,19	10,4	12,6	14,3	15,0	15,1	15,4
		0,28	15,3	18,5	21,1	22,1	22,4	22,7
2	In the descending branch of the thread	0,19	15,4	17,6	19,3	20,0	20,1	20,4
		0,28	20,3	23,5	26,1	27,1	27,4	27,7
With friction radius r=4								
1	On the shuttle's grip	0,19	13,8	16,8	19,0	20,0	20,2	20,6
		0,28	20,4	24,8	28,0	29,4	29,8	30,2
2	In the descending branch of the thread	0,19	18,8	21,8	24,0	25,0	25,2	25,6
		0,28	25,4	29,8	33,0	34,0	34,8	35,2

Influence of the angle and friction coefficient on the thread tension when the thread moves on a stationary cylinder. When the friction radius r=1.

№	Tension Utochiny, CH	Coefficient of friction of the thread on the hook grip, f	Angle of friction of the sink against the gripper, $^{\alpha}$, deg.					
			116	141	159	166	169	172
1	On the shuttle's grip	0,19	5,6	9,7	12,2	13,0	13,4	13,7
		0,28	7,9	12,6	15,7	16,9	17,4	17,8
2	In the descending branch of the thread	0,19	10,6	14,7	17,2	18,0	18,4	18,7
		0,28	12,9	17,6	20,7	21,9	22,4	22,8
At friction radius r=2								
1	On the shuttle's grip	0,19	11,1	19,4	24,4	26,1	26,7	27,3
		0,28	15,7	24,7	31,4	33,	34,7	35,6
2	In the descending branch of the thread	0,19	16,1	24,4	29,4	31,1	31,7	32,3
		0,28	20,7	29,7	36,4	38,8	39,7	40,6

With friction radius r=3

№	Tension Utochiny, CH								
1	On the shuttle's grip	0,19	16,7	29,2	36,7	39,2	40,2	41,1
		0,28	23,6	37,1	47,2	50,8	52,2	53,5
2	In the descending branch of the thread	0,19	21,7	34,2	41,7	44,2	45,2	46,1
		0,28	28,6	32,1	52,2	55,8	57,2	58,5

With friction radius r=4

№	Tension Utochiny, CH								
1	On the shuttle's grip	0,19	22,2	38,8	48,8	52,2	53,4	54,6
		0,28	31,4	49,4	62,8	67,6	69,4	71,2
2	In the descending branch of the thread	0,19	27,2	43,8	53,8	57,2	58,4	59,6
		0,28	36,4	54,4	67,8	72,6	74,4	76,2

Table 3.5.

Influence of the angle and friction coefficient on the thread tension when the thread moves on a rotating cylinder. When the friction radius r=1.

№	Tension Utochiny, CH	Coefficient of friction of the thread against the hook grip, f	Angle of friction of the sink against the gripper, $^{\alpha}$, deg.					
			116	141	159	166	169	172
1	On the shuttle's grip	0,04	4,1	7,4	8,9	9,4	9,5	9,6
2	In the descending branch of the thread	0,04	9,1	12,4	13,9	14,4	14,5	14,6

At friction radius r=2

№	Tension Utochiny, CH		116	141	159	166	169	172
1	On the shuttle's grip	0,04	8,2	14,7	17,8	18,7	19,0	19,2
2	In the descending branch of the thread	0,04	13,2	19,7	22,8	23,7	24,0	24,2

With friction radius r=3

№	Tension Utochiny, CH		116	141	159	166	169	172
1	On the shuttle's grip	0,04	12,3	22,2	26,7	28,2	28,5	28,8
2	In the descending branch of the thread	0,04	17,3	27,2	31,7	33,2	33,5	33,8

With friction radius r=4

№	Tension Utochiny, CH		116	141	159	166	169	172
1	On the shuttle's grip	0,04	16,4	29,6	35,6	37,6	38,0	38,4
2	In the	0,04	21,4	34,6	40,6	42,6	43,0	43,4

descending branch of the thread			.					

To carry out the experiment (Fig.3.6) we used guides 1 with different diameters (cylinders), thread 5 and a spring-loaded, portable strain gauge 2. The method of performing the experiment is as follows. The thread 5 was thrown over the guiding organs 1, a weight was suspended on one end of the thread, and the other end of the thread was suspended on the lever of strain gauge 2, which was uniformly moved by knob 4 and brought the thread to the state of uniform sliding on the cylinder (guiding organ). On the scale of the strain gauge the magnitude of the load at the moment of uniform sliding of the thread was noted. The strain gauge reading corresponds to the tension of the leading branch of the tested thread (K), and the value of the load suspended at the other end of the thread corresponds to the tension of the slave branch (Ko).

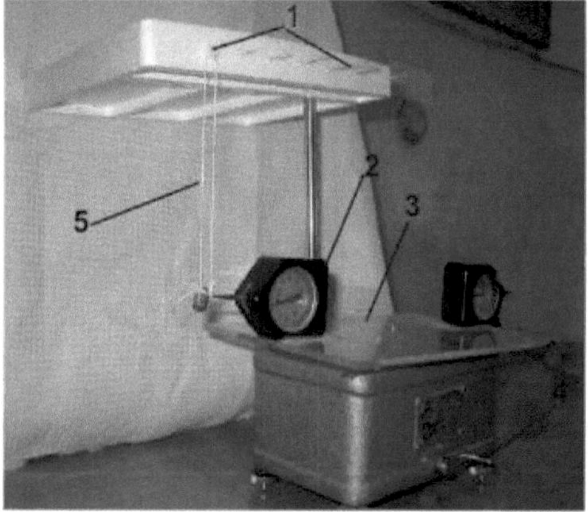

Fig.3.6.

The value of the friction coefficient was calculated on the basis of L. Euler's formula:

$$f = \frac{\lg K - \lg K_o}{\alpha \cdot \lg e}$$

The experiments were carried out at a constant friction angle equal to 180^0, i.e. $\alpha = \pi$ Table 3.6 shows the results of calculation of the friction coefficient for different friction surfaces at $C_{o}=5gr$ and $\alpha=\pi$, yarn linear density 320 tex. Table 3.6 shows that at the beginning of yarn sliding the surface corresponding to the friction coefficient at rest in all variants of the experiment is greater than the values of the friction coefficient in motion corresponding to the state of the friction pair (yarn-surface) of uniform motion. The smallest value of the friction coefficient when using the surface of the rotating cylinder, and the largest when using the rubber surface on the stationary cylinder.

Table 3.6.

Results of friction coefficient calculation for different friction surfaces.

№	Friction surface	Condition of the friction pair	Strain gauge reading, gr	Coefficient of friction of the thread against the surface
1	Rotating cylinder, plastic-metal surface	uniform motion	5,5	0,04
		at the start of movement, at rest	6,0	0,07
2	Flat, fixed cylinder surface plastic-metal surface	uniform motion	8,0	0,19
		at the start of movement, at rest	8,7	0,23
3	Flat, fixed cylinder surface plastic-metal surface	uniform motion	10,0	0,28
		at the start of movement, at rest	15,0	0,45
4	Flat, stationary cylinder surface rubber	uniform motion	20,0	0,56
		at the start of movement, at rest	25,0	0,66

In summary we note: laws of movement of shuttle-capture in the shed have been obtained; equations of tension of weft thread sliding along the plane, along the circumference of fixed and movable cylinders have been obtained, taking into account stiffness of the weft, radius, angle and friction coefficient; it is expedient to use movable cylinder as a gripper in the shuttle; values of the angle of friction of the sinker against the grip depending on the position of the shuttle-gripper in the shed are determined; a stand and a method of determination of the coefficient of friction of the sinker against the grip of the shuttle depending on the shape, sizes and states of the friction surfaces of the grip of the shuttle are developed; increase in the radius of friction of the thread against the grip of the shuttle leads to an increase in the tension of the sinker.

Chapter four summarises the physical, mechanical, hygienic and consumer properties of the fabrics and shows the structure of the fabric and selvedges.

The structure of the fabric produced on the modernised weaving machine type AT-100 is shown in Fig. 4.1.

Off the reel Off the reel

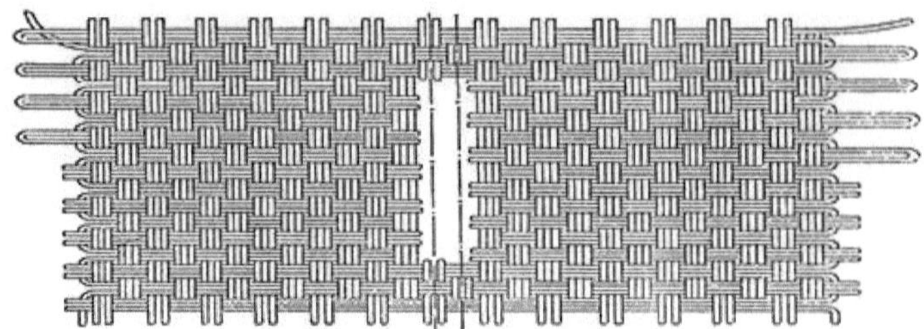

Figure 4.1. Fabric and edge structure

With each movement of the hook, two parallel weft threads are inserted into the shed. The laying takes place alternately - from one and the other bobbin, the threads of which can be the same or different depending on the assortment being produced. The laid double weft from one bobbin ends with a loop dropped from the shuttle catch. This filling yarn is wrapped around the bottom with another weft from another bobbin and also ends in a loop on the opposite side of the fabric from the bobbin. This selvedge structure provides strength and prevents the edge warp threads from weaving out of the fabric. Tables 4.1-4.2 show the physical and mechanical properties of the selvedges of fabrics produced on the basic and modernised weaving machine

Table 4.1.

Physical and mechanical properties of a fabric strip with selvedge.

№	Samples	Fabric density		Breaking load, in kG	Elongation in %	Yarn yield in %		Breaking force per 1 cm width in kg
		On the basis of	On the duck			On the basis of	On the duck	
1	With plain edge	87	90	39,0	18,0	17,5	6,1	7,9
2	With edge produced on a modernised machine	87	92	38,3	15,5	15,6	3,7	7,86

Shear test results of the fabric.

№	Samples	Applied force in kg	Number of threads shifted	Single thread force in kg	Result
1	Fabric background	4,3	8	0,56	Thread shift
2	Fringed edge	33,2	-	-	Rupture
3	Edge with trimmed fringe	30,2	-	-	Rupture

Table 4.1 shows that the selvedge of the fabric produced on a weaving machine with a fixed weft bundle can withstand the same tearing forces as the selvedge of a fabric produced on a conventional weaving machine. Table 4.2 shows that in the edge shear test (main yarns spreading) no shear occurs but a tear is

obtained. The magnitude of forces for shearing of the main threads in the selvedge is 7-8 times higher than the forces required for shearing of the threads in the background. Also in the work the fabric properties were worked out and determined. Three samples of fabrics were worked out, where the main yarn of linear density T=280 tex was used in the first variant in two folds **T=280** texx2, in three folds T=280 **tex3**, in four folds T=280 **tex4**, and other fabric parameters remained unchanged. Table 4.3 shows the filling calculation of fabrics.

Table 4.3.

Fabric Filling Calculation.

№	Name	Unit.	Fabric production options		
			1	2	3
1	Reed width	cm	100	100	100
2	Linear density of the warp	Tex	280/2	280/3	280/4
3	Linear weft density	Tex	320/2	320/3	320/4
4	Fabric density on the base	n/dm	40	40	40
5	Weft density	n/dm	40	40	40
6	Number of threads in the warp	thread	400	400	400
7	Number of remises in a refuelling	piece	2	2	2
8	Reed number	Tooth/dm	40	40	40
9	Surface density of the fabric	Gr/m^2	545	678	805
10	Workmanship on the basis of	%	18,6	16,3	13,1
11	Finishing in weft	%	5,6	8,0	11,2

Table 4.4.

Machine threading parameters for new fabric production.

№	Name of parameters	Unit.	indicators
1	Position of the scalo relative to the sternum	mm	+30
2	Reed height	mm	70
3	Anterior pharynx	mm	160
4	Posterior pharynx	mm	440
5	Margin value	mm	40

The threading parameters for the new fabric are shown in Table 4.4. In the modernised weaving machine system, each flip of the shuttle-gripper corresponds to the insertion of two weft threads into the shed, ending in loops behind the fabric edge, which after formation (see Fig. 4.1) go into the wefts. Weft grades are determined

$$Y = \frac{l}{B + l} \cdot 100\% \qquad\qquad (1)$$

where: l *is the* length of the weft loop; B is the width of the fabric tuck on the reed.

Equation (1) shows that the larger the reed filling width of the fabric, the smaller the foreshortening and vice versa. The length of the weft loop on each edge is 30-50mm. Therefore, with a working width of 100cm on the reed for both edges of the fabric, the foreshortening will be:

$$Y = \left[\frac{30 \cdot 2}{1000 + 60} \cdots \frac{50 \cdot 2}{1000 + 100} \right] \cdot 100 = 5{,}6 \div 9\%$$

Table 4.5 shows load-elongation diagrams for warp yarns of linear densities 280tex, 280texx2, 280texx3, and 280texx4 based on load-elongation diagrams, and Fig.4.3 shows load-elongation diagrams for weft yarns of linear densities 320tex, 320texx2, 320texx3, and 320texx4..3 load-elongation diagrams for weft yarns of linear density 320tex, 320texx2, 320texx3, and 320texx4, obtained on computerised tensile machines with a tensile strength of 1000 N, show the results of breaking load and elongation tests.

Table 4.5.

Results of breaking load and elongation tests on yarns

Conventional machine		Modernised machine	
To=280x2			
Basis			
P	*l*	*P*	*l*
106	18	112	20
93,5	20	112	20
96	20	112	18
Duck			
118	25	124	20
89	22	113	20
79	15	128	20
To=280x3			
Basis			
146	45	178	35
158	40	171	30
214	40	233	35
Duck			
124	20	156	20
118	20	132	20
111	20	114	20
To=280x4			
Basis			

167	30	195	30
209	28	221	30
151	30	168	33
Duck			
108	25	124	22
101	25	135	20
134	25	134	20

The analysis shows that the weft yarns are almost twice as strong as the warp yarns in terms of breaking properties, with a 12.5 per cent increase in the linear density of the weft yarn in relation to the warp yarn. The fabric produced on the modernised weaving machine differs in appearance from the two-strand fabric produced on the conventional weaving machine only by the fact that a fringe of weft loops is formed in the selvedges of the fabric. The physical and mechanical properties of the fabrics produced on a weaving machine with a fixed weft bundle and on a conventional weaving machine were compared. The results of the fabric test are summarised in Table 4.6. The table shows that the strength of the main strip of the fabric produced on the conventional machine and the modernised machine are the same. The strength of the weft strip from the conventional machine is higher. This is explained by the fact that the weft yarn on the modernised machine experiences a jerk at the moment of its pick-up, rubs against the guides during its movement, etc. As a result, the weft is somewhat deformed and its strength and elongation are reduced. The weft elongation of a fabric made on a modernised machine is less than that of a fabric made on a conventional machine.

Table 4.6.

The results of the fabric test.

The fabric from the loom	Linear density threads, tex		Number of threads per 50 mm		Value of indicators when testing a strip of fabric 50*200 mm				Strength per 1 strand in CH	
					on the basis of read-udley-competence kG. cmm.		by duck strength, kG.s	elongation mm.		
	basis	ducks	basis	ducks					basis	ducks
Modernised 560	640	20	20		112	19,3	121,6	20	5,6	6,08
Ordinary 560	640	20	20		98	19,3	95	20,6	4,9	4,75
Modernised 840	640	20	20		194	33,3	134	20	9,7	6,7
Ordinary 840	640	20	20		172	41,6	117,6	20	8,6	5,88
Modernised 1120	640	20	20		194,6	31	131	20,6	9,73	6,55
Ordinary 1120	640	20	20		175,6	29,3	114,3	25	8,78	5,7

In the fifth chapter structure parameters and properties of fire hoses are given, elements of fire hose structure are considered, fire hose structure indicators are considered. Fire hose is a flexible pipeline for transporting fire extinguishing substances equipped with fire connecting heads. Fire hoses are made of tarpaulin impregnated with a special composition. tarpaulin or synthetic fabric and are designed for a working pressure of at least 1.0 MPa. MPa. To increase water resistance, durability and protection against aggressive media (oil products, acids, high and low temperatures) fire hoses can have rubber Fire hoses may have rubber or polymer coating inside and metal reinforcement (braid) or polymer coating outside. fire hoses are subdivided into receiving and outgoing hoses. The former are designed for water intake from a reservoir or water supply, and the latter are used to direct water from the pump, creating the necessary pressure, to the burning objects through the outlet trunks (fire sprays). Receiving hoses are manufactured at rubber industry plants. Outlet hoses are products of the textile industry. Fire hoses have the following basic requirements: high tensile strength, water resistance, resistance to microorganisms and friction, as well as good flexibility and low weight. The tensile strength, i.e. the resistance to bursting, must correspond to the water pressure developed by the fire pump. The watertightness of fire hoses allows the water pressure generated by the pump to be maintained along the entire length of the hose during operation.

Fire hoses often remain wet for a long time after use, which creates favourable conditions for the development of various microorganisms. In this case, the moisture remains on the surface of the hose for a long time and inside the hose for a longer period of time even in warm sunny weather. Therefore, resistance to microorganisms is one of the important properties of fire hoses. Very often in the process of extinguishing the fire hoses filled with water, dragged across the asphalt, stones or on the ground, as well as through window openings on broken glass, often they come into contact with roofing iron. In these conditions it is necessary to have hoses resistant to abrasion and external damage. Outlet sleeves are divided into rubberised and non-rubberised (linen). In small quantities, there are also made comb sleeves designed to work at low water pressure. The maximum use of yarn strength in the fabric and the tightest binding of warp with weft to obtain a dense, high-strength, waterproof, resistant to external mechanical influences of the fabric provides plain weave. Therefore, this type of weave is used in the production of non-rubberised fire hoses (i.e. linen). Hoses that are rubberised, i.e. sheaths, are also produced by plain weave. The sheaths need only have a high tensile strength, and water resistance is achieved by the introduction of a rubber chamber. A rubberised hose consists of a solid woven textile hose, called a sheath, and a rubber chamber glued into it to seal the walls. Full sealing of the hose walls is achieved by gluing rubber chambers inside. Covers are mainly made of cotton yarn in the base and linen yarn in the weft (semi-linen) and in small quantities of cotton yarn in the base and capron yarn in the weft. When making filling calculations for the production of half-linen covers on flat weaving machines, it is necessary to know the strength of the sleeve walls in the warp and weft. The strength of the sleeve walls must correspond to the hydraulic pressure for which the sleeve is designed. Depending on the hydraulic pressure they can withstand, sleeves are divided into three groups: normal, reinforced and high-strength sleeves. Sleeves are produced with coloured wires in each strand along the entire length of the sleeve. Normal - with one webbing, reinforced - with two webbing and high-strength sleeves - with three webbing. Linen sleeves used without slitting are waterproof due to the ability of the linen yarn to swell quickly when wet and close the pores (spaces between the yarns) of the fabric. Swelling can only sufficiently waterproof a sleeve if the pore size is minimised. Therefore, the fabric density, and hence the filling in the warp and weft should be maximised (around 100%). In addition, it has been found that it is possible to produce linen sleeves with satisfactory water resistance if yarns of uniform thickness are used. Dry spun yarns are used for linen sleeves as they have the greatest ability to swell quickly when wet. The weight distribution of yarns in linen sleeves is usually as follows: 60 per cent main yarn and 40 per cent weft yarn. The tension in the sleeve walls under hydraulic pressure in the warp is 2 times lower than in the weft. The question therefore arises as to the rationality of this sleeve structure. However, the discrepancy between the number of warp yarns and the number of weft yarns is justified by the following circumstances. In order to obtain the necessary water resistance, the main yarns must be arranged with the highest possible density (at a filling close to 100 per cent), and the pores in the fabric must disappear when the yarns get wet and therefore swell. In such sleeves, the weft should have a minimum elongation, i.e. be located in the fabric in a straightened state, as in this case it seems to fulfil the role of hoops, contributing to the compaction of the main threads when they swell. This explains the too elastic twist of the weft yarn, which ensures minimum yarn elongation. Linen fire hoses have a standard, which includes specifications, filling calculations, instructions for disassembly of hoses, acceptance rules and test method, packaging and labelling rules, as well as storage and transport conditions. Linen hoses depending on the value of the applied hydraulic pressure are divided into the following groups: light, normal and reinforced. Linen hoses are produced with an inner diameter of four sizes: 26, 51, 66 and 77 mm.

The hoses shall withstand the hydraulic pressure specified in Table 5.1.

Hydraulic pressure of fire hoses

Inner diameter in mm	Working hydraulic pressure in kg/cm² in hoses			Working hydraulic pressure in kg/cm² when testing hoses		
	normal	reinforced	higher strength	normal	reinforced	higher strength
51	12	14	16	15	18	20
66	12	14	16	15	18	20
77	12	14	-	14	16	-
89	-	12	-	-	16	-

Table 5.2 summarises sleeves produced on flat weaving machines and Table 5.3 summarises sleeves produced on round weaving machines.

Table 5.2.

Sleeves produced on flat weaving machines

Inner diameter of hoses in mm	Number of cotton basic yarns	Number of main threads	Linen weft yarn number	Number of weft yarns per 10 cm	Synthetic weft yarn number	Number of weft yarns per 10 cm	All 100 m of hose in kg, not more	Intertwining
Rubberised sleeves are normal								
51	27/9	321±4	6/14	34± 2	-	-	71,6	linen
66	27/9	417±5	6/16	34± 2	-	-	99,8	linen
77	27/9	524±5	6/18	34± 2	-	-	109,4	linen
Reinforced rubberised sleeves								
51	27/9	321±4	6/16	34± 2	-	-	74,8	linen
66	27/9	460±5	6/18	34± 2	34/50	38± 2	96,9:86,5	linen
77	27/9	572±5	6/20	34± 2	34/50	38± 2	116,2-105	linen
89	27/9	700±6	-	34± 2	34/50	38± 2	121,4	linen
Heavy duty rubberised sleeves								
51	27/9	364±4	6/18	34± 2	-	-	78,2	linen
66	27/9	524±5	-	-	34/50	38± 2	91,1	linen

Sleeves produced on circular weaving machines

Inner diameter of hoses in mm	Number of cotton basic yarns	Number of main threads	Linen weft yarn number	Number of weft yarns per 10 cm	All 100 m of hose in kg, not more	Intertwining
Rubberised sleeves are normal						
51	27/9	324±4	6/10	34± 2	68,4	linen
66	27/9	420±5	6/14	34± 2	88,68	linen
77	27/9	528±5	6/14	34± 2	104,7	linen
Reinforced rubberised sleeves						
51	27/9	324±4	6/14	34± 2	74,1	linen
66	27/9	468±5	6/14	34± 2	93,2	linen
77	27/9	576±5	6/16	34± 2	113,2	linen
89	27/9	708±6	6/18	34± 2	134	linen
Heavy duty rubberised sleeves						
51	27/9	360±4	6/16	34± 2	76,8	linen
66	27/9	528±5	6/16	34± 2	97,7	linen

The peculiarity of double-layer fabrics is that two independent warp and two independent weft systems are involved in their construction. At the same time they form two independent layers lying one on top of the other, free or connected to each other in the weaving process. Sack (hollow) weaves are used for the production of fire or field hoses, technical cloth, sacks. The basic weaves are plain weave, corduroy 2/2, rep weft 2/2, four-strand sateen. The following principles are used in the formation of sack (hollow) weaves:

1 .The warp and weft threads of the upper fabric layer are numbered with Arabic numerals and the warp and weft threads of the lower fabric layer with Roman numerals.

2 .When depicting a weave on paper, the warp and weft yarns are conventionally shifted to the same plane.

3 .When weft is brought into the top layer, all warp threads of the bottom layer are dropped (Fig. 5.1.a).

4 .When weft is added to the bottom layer, all warp yarns of the top layer are raised (Fig. 5.1.b).

5 The lifting of the warp yarns of the upper layer when weft is inserted into the lower layer of the fabric is shown in the drawing as a circle.

6 The layers are joined at the edges of the filling by alternate laying of the

weft.

7   The warp and weft rapport (R) is equal to the smallest multiple of the number of yarns in the rapport (R_6) of the base weave multiplied by the number of layers (K).

$$R = R_6 \cdot K \qquad\qquad (5.1)$$

8   .Bass cords are lowered when forming the upper layer of the hollow fabric, and raised when forming the lower layer of the fabric (Fig.2.2)

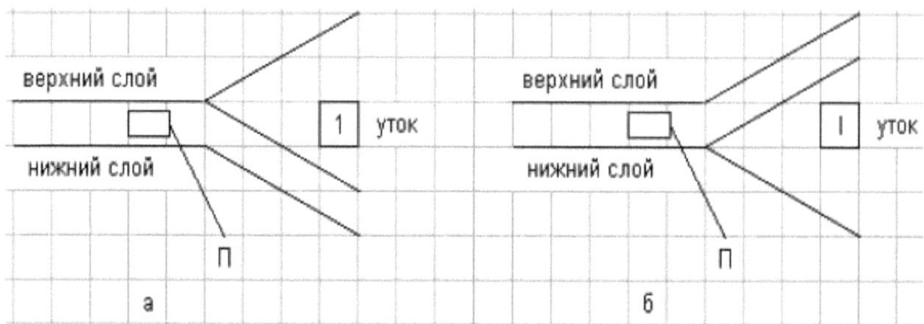

Fig. 5.1. Weft laying in the formation of a (baggy) hollow fabric.

To maintain the established density of the hollow fabric, bass cords are used in the places of transition of the weft from one layer to another, i.e. in the places of folds of the hollow fabric. The cords are threaded into individual wefts and reed tines.

The bass cords are lowered during the formation of the upper fabric layer and raised during the formation of the lower fabric layer. This means that they are not caught by the weft (Fig. 2) and can be freely removed after removing the fabric from the loom.

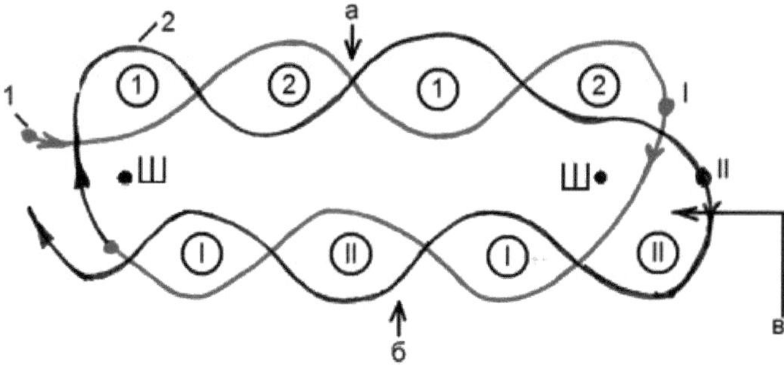

Fig. 5.2. Internal weave of the bottom layer of the web (bag) hollow fabric.

Let's construct the filling pattern of a hollow fabric on the basis of plain weave, the rapport of the basic weave $R_6 = 2$, and the ratio of fabric layers we will accept 1:1, (Fig. 5.3. a and 5.3. b).

Hollow fabric warp and weft ratio

$$R_o = R_y = 2R_6 = 4 \qquad (5.2)$$

Draw the cross section of the fabric (Fig.5.4) and determine the internal weave of the bottom layer of the fabric (Fig.5.1 and 5.3.c). We then transfer the weave of Fig.5.3.a and Fig.5.4.c to the filling pattern 5.5. Using the principles of points 3, 4, 5 and 8 of the hollow fabric construction, we draw up the complete filling pattern of the hollow fabric. The alternation of weft skeins takes place in the following sequence:

-   first weft laying for the top layer;
-   second weft laying for the bottom layer;
-   third weft laying for the top layer;
-   fourth weft laying for the bottom layer.

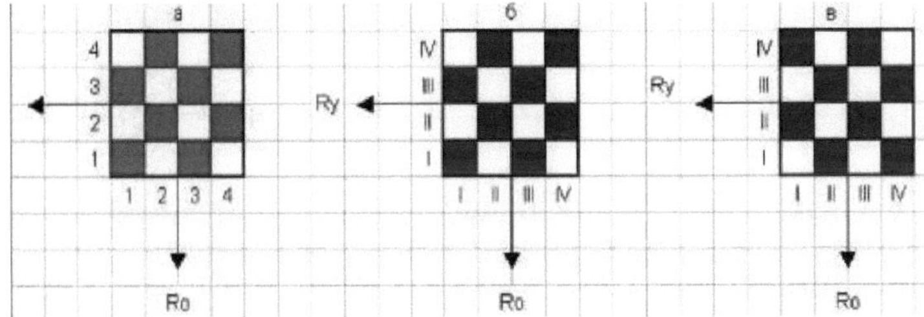

Fig.5.3. Filling pattern of (sack) hollow fabric: a - top layer; b - bottom layer; c - inner weave of the bottom layer of the fabric

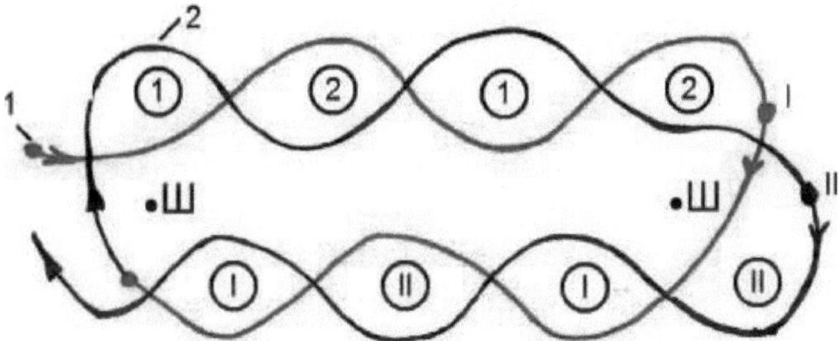

Figure 5.4: Transverse section of (sac) hollow tissue.

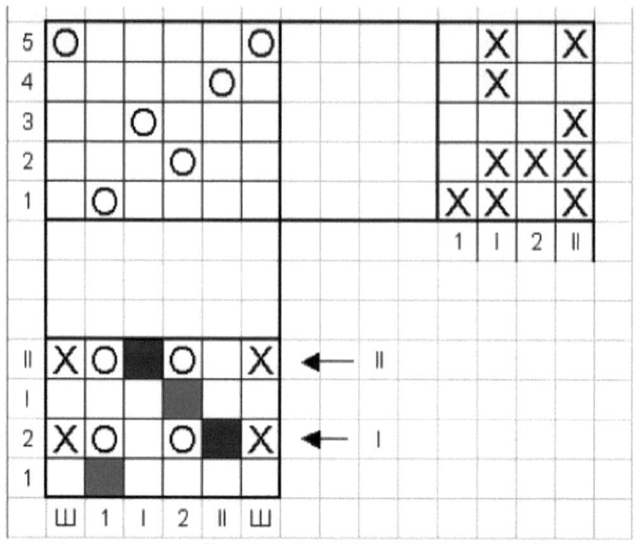

Figure 5.5. Complete filling pattern of a hollow fabric

The bass cords (B) are lifted from the individual loom when the bottom layer of the fabric is formed. The process of removing bass cords after removing the hollow fabric from the loom is very labour-intensive. Therefore, it is expedient to develop means of maintaining a given density of the fabric in the places of folding (transition of wefts from one layer to another layer), which replaced the bass cords on the machine. One of the variants of location between the layers in the fabric at the fabric edge is a bar (P) - rectangular or trapezoidal, oval and other shapes (Fig.5.1). Indicators of the structure of fire hoses determine the dependencies between the parameters of their structure and physical and mechanical properties. It is necessary to take into account the purpose of the fabric, the properties of the raw materials used, the features of the technological process and the formation of textile materials. Pressure (discharge) hoses are used as flexible pipelines for supplying water under pressure, as well as for pumping fuel, fuel oil and oils. One of the main requirements for outgoing fire hoses is water resistance, which allows the water pressure generated by the pump to be maintained along the entire length of the hose. The water resistance of hoses made of natural fibres (cotton, linen) is achieved due to the ability of linen yarn to swell quickly when wet and close the gaps between the threads in the fabric. Linen sleeves are usually produced on flat looms in folded form and therefore have low weft strength at the selvedges. The weight of the main yarn in the sleeve is about 60 per cent and the weft only 40 per cent, while the stress in the sleeve walls under hydraulic pressure in the warp direction is half that in the weft direction. This discrepancy between the sleeve structure and the requirements for its strength in the warp and weft directions is caused by the need to ensure maximum filling of the fabric in the warp direction and minimum elongation of weft yarns under the action of hydraulic pressure. More rational is the structure of sleeves and covers produced on circular weaving machines. When calculating the structure parameters of a round woven sleeve, the forces acting in the axial direction (along the warp yarns) are determined according to the formula:

$$Q_o = \frac{cpD^2 3z_o}{4K_o} \qquad (5.3)$$

where c - hydraulic pressure, kgf/cm^2 ; D - average diameter of the hose, cm; z - correction factor, taking into account the increase in the diameter of the hose under the action of hydraulic pressure; K_o - coefficient of using the strength of the warp yarn; z_o - safety factor of warp strength.

The value of the tangential forces acting in the circumferential direction on the weft yarns is calculated according to the formula

$$Q_y = \frac{pD\eta_1 10\eta_2 z_y}{2K_y} \qquad (5.4)$$

where η_2 is the coefficient that takes into account the increment of the sleeve length (L=10 cm); K_y is the weft yarn strength utilisation factor; z_y is the weft safety factor.

The calculated diameter of the sleeve cover is determined based on the given value of the inner diameter of the sleeve DB, the wall thickness of the waterproofing layer Δ and the diameters of warp d_o and weft d_y.

$$D = DB + 2\Delta + 2\, d_o + d_y \qquad (5.5)$$

For approximate calculations of the strength of the cover fabric on the base (50 mm wide strip) we find the following expression from the formula (5.3):

$$Q_o' = \frac{pDz_o}{0{,}8K_o} \qquad (5.6)$$

From formula (5.4) we obtain the strength of the fabric strip in the weft:

$$Q_y' = \frac{pDz_y}{0{,}4K_y} \qquad (5.7)$$

From the analysis of experimental data it was found that for sleeve covers produced on a round weaving machine, it is possible to take the coefficient $K°{=}0.5$ and $Ku{=}0.7$. The strength reserve of the sheath walls in the warp direction is 1.5, and in the weft direction - 1.25. Increasing the safety margin in the base is caused by the fact that when moving and bending the sleeve filled with water under pressure, the main threads experience greater stress than the weft, which they at the same time protect from external damage. Based on the required warp strength of the strip and the breaking load of a single yarn, the filling number of warp yarns around the entire circumference of the cover is calculated.

$$m_o = \frac{\pi D Q_o'}{5 p_o m_1} \qquad (5.8)$$

Weft density of the cover fabric per 10 cm

$$S_y = \frac{2 Q_y'}{p_y m_1} \qquad (5.9)$$

where py is the breaking load of a single thread, kgf; $m1$ is the number of torsion folds.

The percentage of linear filling of the sleeve cover fabric is calculated using a known formula:

$$3_o...3_o + 3_y - 0,013_o3_y \qquad (5.10)$$

where zo and zu are the percentage of linear filling in the warp and weft.

It is recommended to determine the value of zo as the product of the strand diameter dn by the strand density $\frac{S_o}{m^2}$, where $m2$ is the number of strands tucked into the galley. At $t2=2\ m^2$ the strand diameter will be equal to

$$d_n = 0,0357 \sqrt{\frac{2T}{\gamma_o}} \qquad (5.11)$$

where To is the warp thickness, tex; γ_o is the volumetric weight of the yarn, g/cm^3.

The weight of 1 pg. m (surface density) of the sleeve cover can be calculated using the formula:

$$M = \frac{m_o T_o}{10^3 (1 - 0,01 a_o)} + \frac{S_y \pi D T_y}{104 (1 - 0,01 a_y)} \qquad (5.12)$$

where ao and Au are the degree of warp and weft processing, %.

Table 5.4 shows the structure parameters of fire hoses with a diameter of 67-68 mm.

Table 5.4.

Structure parameters of fire hoses with a diameter of 67-68 mm.

Indicators		Reinforced linen sleeve
Thread thickness, tex	fundamentals	167x2
	duck	167x10

Breaking load of yarns, kgf	fundamentals	6,95
	duck	36,6
Number of warp threads per circumference		592
Yarn density per 10 cm in weft		42
Yarn processing degree, %	on the basis of	19,3
	on duck	6,6
Linear density of the cover, ktex		399
Fabric filling, %	on the basis of	100
	on duck	63,5

The design of fabric with given properties is also carried out. For the correct construction of the sleeve fabric it is necessary to know about the stresses arising in it during operation, about the technology of manufacturing rubber sleeves, about the distribution of forces appearing as a result of internal pressures at which the sleeves are in working condition. For example, let us take a simple sleeve consisting of an inner rubber layer, rubberised fabric gaskets and a rubber lining. When assembling the sleeve inner rubber layer, which is a rubber tube, stretched on a metal rod (mandrel) of a certain diameter, the layer of rubber wrapped with a strip of rubberised fabric, which is then carefully rolled. Rolling is done in order to ensure tight adhesion of the fabric to the rubber and individual layers to each other. The rubber strip, which forms the outer lining of the sleeve, is applied last and is also well rolled. The assembled sleeve is wrapped tightly with cloth bandages and vulcanised. After vulcanisation, the bandages are removed. Let us consider the stresses to which the sleeve walls are subjected in different directions at a certain internal pressure. Let us denote by P *the* pressure per $1 cm^2$, available inside a sleeve of radius r. The walls of the sleeve have sufficient resistance to balance the internal pressure P. Let us determine the force that tends to break the sleeve along the circumference ***tpor*** (Fig. 5.6), and the force that tends to break it along the formers AD and BC. It will be seen from the figure that any other plane which divides the sleeve into two parts at a different angle will give a line of greater length at the intersection with the walls. Thus, two cases are taken in which the force for the minimum length of the sleeve will be maximised.

The pressure experienced by the arm along its axis will be equal to

$$P\pi r^2 \qquad (5.13)$$

where r is the radius of the sleeve in cm.

The length of the circumference of ***tpor***, representing the rupture line of *the* sleeve wall, is *2nr*.

Hence the axial wall stress per 1cm length will be:

$$T_0 = \frac{P\pi r^2}{2\pi r} = \frac{1}{2}Pr \qquad\qquad (5.14)$$

Let the length of the sleeve element be *l*. The surface of both half-cylinders is subjected in each of its points to pressure *P*. *It* is known that the projection of the surface of the two half-cylinders on the plane *ACBD* is equal to the area of this rectangle.

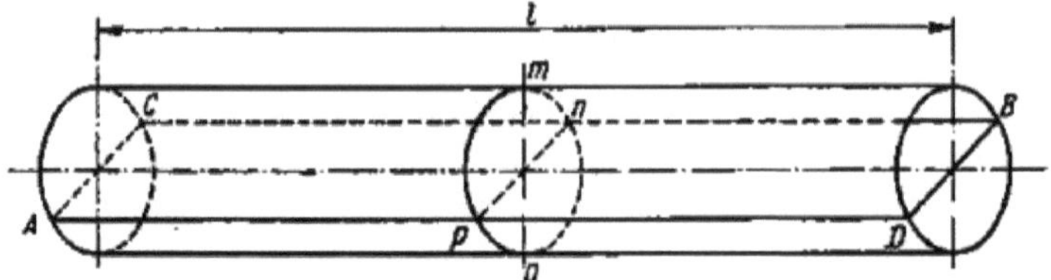

Figure 5.6. Notional representation of the sleeve.

The pressure per area of the rectangle is equal to *2Plr* and acts on formers *AB* and *CD*, whose length is equal to *2l*. From this we can conclude that the cross-sectional area of the wall of the rubberised sleeve must be twice as large as the cross-sectional area of the sleeve formers. If these cross-sectional areas are equal, the fabric of which the sleeve is made must have a resistance 2 times greater in the radial direction than in the direction of the sleeve axis. This fact must be taken into account when wrapping it with fabric. Sleeve fabrics should be made with such a difference of some indicators in the stern, so that after rubberising this difference disappeared altogether or had the smallest value. Therefore, the weft strength of the sleeve fabric should be greater than the warp strength, and the weft elongation should be less than the warp elongation. Water permeability of the fabric was calculated and analysed. Water permeability characterises the ability of products to pass water through themselves. The water permeability characteristic is the water permeability coefficient, which is expressed by the amount of water V (dm) passing through 1 m^2 of the fabric surface for 1 s at liquid pressure g, Pa

$$B = \frac{V}{F \cdot t} \qquad \frac{\partial \text{м}^3}{\text{м}^2 c} \qquad\qquad (5.15)$$

where V is the volume of water that has passed through the fabric of area F during time T. The area tkni is constant equal to 0.04m $.^2$

The volume of water is calculated using the following formula: *m*

$$V = \frac{m}{\rho_{\text{воды}}} \qquad \partial \text{м}^3 \qquad\qquad (5.16)$$

where $p_{\text{воды}}$ is the density of water which is equal to 1g/cm^3 ; *m is the* mass of the sample which is equal to $m = m_n - m_l$, m_n *is the* mass of the sample at a certain time (minutes), m_l *is the* mass of the initial sample.

Water permeability depends on the thickness of the product, its porosity, fibre composition and type of finish. Water resistance is the resistance of textile products to water seeping through them. The minimum water pressure on the test piece that causes the appearance of a third drop of liquid on the opposite surface of the piece is taken as the water resistance value. This characteristic is determined on special devices called penetrometers. Sometimes the method of "koshel" is used, in this case water is poured into the fabric, which is fixed in the form of a bag, up to a height H, and water resistance is characterised by the time after which the third drop of water or a certain volume of water seeps out. To determine the water permeability through a sample of material passing 0.5 dm^3 of water at a temperature of 20 ° C and a constant pressure of 500 n/m^2 , using a stopwatch to note the time for which the specified amount of water passes through the sample. In addition to the pressure at which the moisture is forced through the sample, the thickness and filling of the material influence the water permeability. Water permeability and water resistance are characterised by the time during which the material does not become wet by holding water under constant pressure. Water resistance and water permeability can also be characterised by the lowest pressure at which water penetrates the material. Instruments called penetrometers operate on this principle. Experimental studies have been carried out to determine water resistance in fire hoses (Table 5.5).

Table 5.5.

Experimental studies of fabric water resistance

Water resistance of 2x2 samples								
minutes	cotton				Len				retention time
	sample 1	sample 2	sample 3	sample 4	sample 1	sample 2	sample 3	sample 4	
0	282	268	276	274	236	246	238	240	0
5	316	300	314	312	288	300	288	280	5
10	334	318	328	326	302	326	304	298	5
15	346	320	346	340	330	350	330	318	4,45
30	394	386	390	396	418	428	400	412	8
45	418	410	412	410	452	480	450	464	5,5
60	424	416	422	420	458	494	468	480	9
75	432	428	434	444	478	508	478	500	8
90	440	436	452	454	490	516	482	504	8
120	444	440	454	464	488	516	486	504	8
150	444	440	454	464	490	518	490	504	8
180	444	440	454	464	490	518	490	504	8

Studies were carried out on water resistance of fabrics made of linen and cotton yarn. On the basis of the data obtained by experimental studies, the graphs showing the changes in water resistance of fabrics depending on the time of water exposure were constructed (Fig. 5.7-5.10).

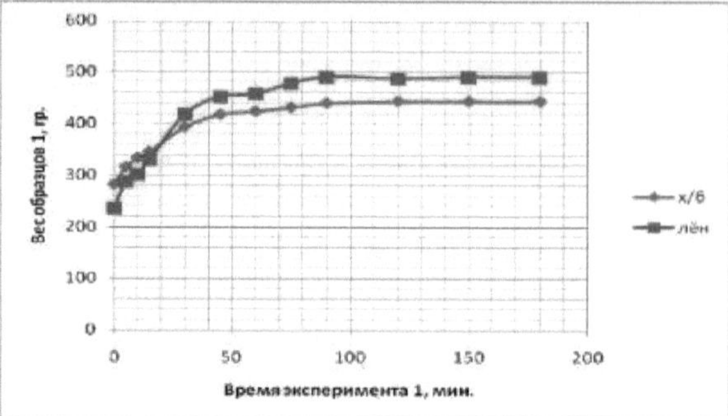

Fig.5.7. Graph of change of water resistance of samples 1 depending on time of water exposure.

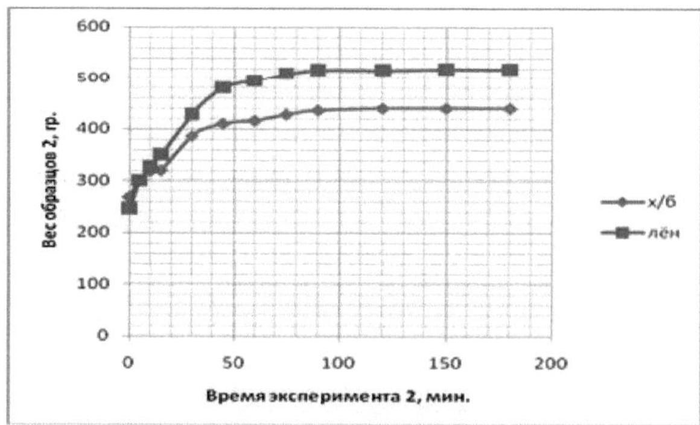

Fig.5.8. Graph of change of water resistance of samples 2 depending on time of water exposure.

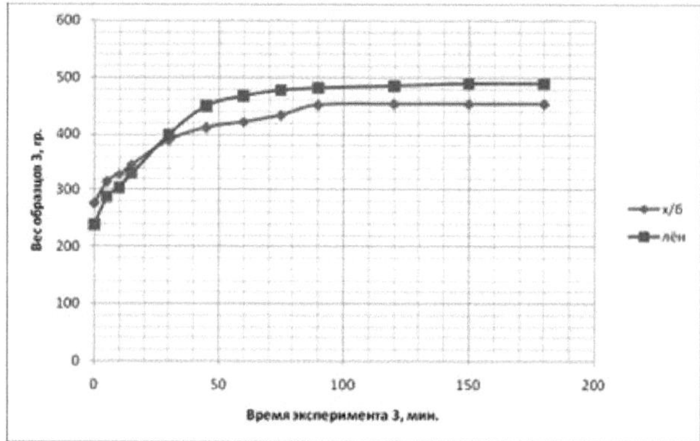

Fig.5.9. Graph of change of water resistance of samples 3 depending on time of

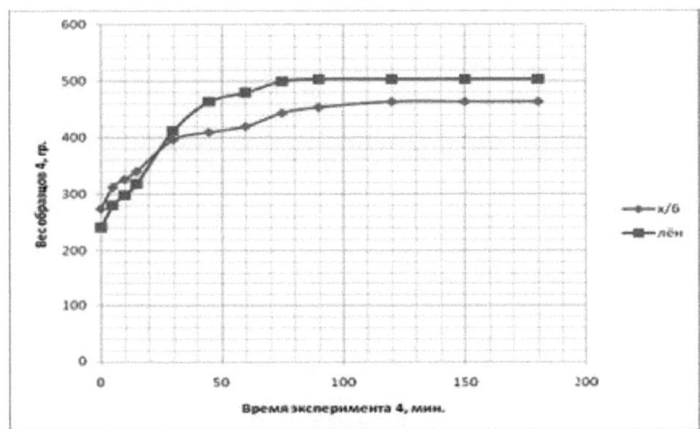

Fig.5.10. Graph of change of water resistance of samples 4 as a function of time of water exposure.

All the data obtained were examined and the averages of the data were calculated and presented in Table 5.6 and a graph was plotted (Figure 5.11).

Table 5.6.

Regularities of change in water resistance of fabrics depending on exposure time

Minutes	cf. cotton	cf. flax
0	240	275
5	289	310,5
10	307,5	326,5
15	332	338
30	414,5	391,5
45	461,5	412,5
60	475	420,5
75	491	434,5
90	498	445,5
120	498,5	450,5
150	500,5	450,5
180	500,5	450,5

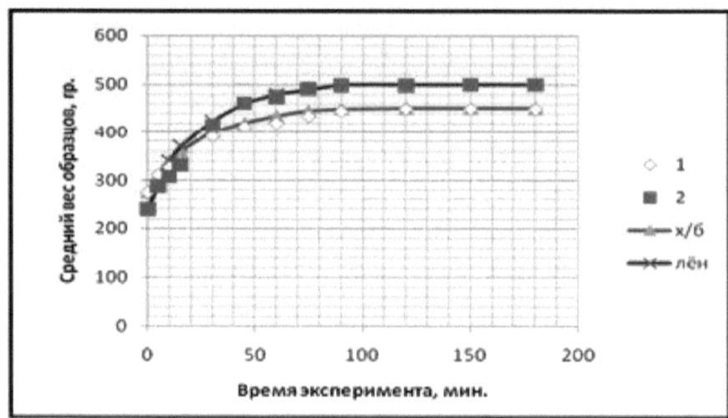

Fig5.11. Graph of changes in water resistance of fabrics depending on the time of water exposure: 1 - experimental parameters of cotton; 2 - experimental parameters of linen

On the basis of the formula (5.13) the calculation of water resistance in fabrics was made and the regularity of change of water resistance of fabrics depending on the time of exposure was established.

Calculating the water permeability of linen fabric in 60 minutes

$$B = \frac{V}{F \cdot t} = \frac{0{,}0235}{0{,}04 \cdot 3600} = \frac{0{,}0235}{144} = 16{,}32 \cdot 10^{-5} \ \frac{\partial м^3}{м^2 c}$$

$$V = \frac{m}{P_{воды}} = \frac{m_n - m_1}{P_{воды}} = \frac{0{,}0475 - 0{,}024}{1} = 0{,}0235 \ \partial м^3$$

$$t = 60 \, м = 60 \cdot 60 = 3600 \ c$$

Calculating the water permeability of linen fabric in 120 minutes.

$$B = \frac{V}{F \cdot t} = \frac{0{,}0258}{0{,}04 \cdot 7200} = \frac{0{,}0258}{288} = 8{,}96 \cdot 10^{-5} \ \frac{\partial м^3}{м^2 c}$$

$$V = \frac{m}{P_{воды}} = \frac{m_n - m_1}{P_{воды}} = \frac{0{,}0498 - 0{,}024}{1} = 0{,}0258 \ \partial м^3$$

$$t = 120 \, м = 120 \cdot 60 = 7200 \ c$$

Calculating the water permeability of linen fabric in 180 minutes.

$$B = \frac{V}{F \cdot t} = \frac{0{,}026}{0{,}04 \cdot 10800} = \frac{0{,}026}{432} = 6{,}02 \cdot 10^{-5} \ \frac{\partial м^3}{м^2 c}$$

$$V = \frac{m}{P_{воды}} = \frac{m_n - m_1}{P_{воды}} = \frac{0{,}05 - 0{,}024}{1} = 0{,}026 \ \partial м^3$$

$$t = 180 \, м = 180 \cdot 60 = 10800 \ c$$

The data of linen fabric water permeability obtained by calculation are entered in Table 10 and a graph is plotted on their basis (Fig.5.11).

Calculating the water permeability of cotton fabric in 60 minutes.

$$B = \frac{V}{F \cdot t} = \frac{0,0145}{0,04 \cdot 3600} = \frac{0,0145}{144} = 10,06 \cdot 10^{-5} \; \frac{\partial \text{м}^3}{\text{м}^2 c}$$

$$V = \frac{m}{p_{\text{воды}}} = \frac{m_n - m_1}{p_{\text{воды}}} = \frac{0,042 - 0,0275}{1} = 0,0145 \; \partial \text{м}^3$$

$$t = 60 \, \text{м} = 60 \cdot 60 = 3600 \; c$$

Calculating the water permeability of cotton fabric in 120 minutes.

$$B = \frac{V}{F \cdot t} = \frac{0,0175}{0,04 \cdot 7200} = \frac{0,0175}{288} = 8,96 \cdot 10^{-5} \; \frac{\partial \text{м}^3}{\text{м}^2 c}$$

$$V = \frac{m}{p_{\text{воды}}} = \frac{m_n - m_1}{p_{\text{воды}}} = \frac{0,045 - 0,0275}{1} = 0,0175 \; \partial \text{м}^3$$

$$t = 120 \, \text{м} = 120 \cdot 60 = 7200 \; c$$

Calculating the water permeability of cotton fabric in 180 minutes.

$$B = \frac{V}{F \cdot t} = \frac{0,0175}{0,04 \cdot 10800} = \frac{0,0175}{432} = 4,05 \cdot 10^{-5} \; \frac{\partial \text{м}^3}{\text{м}^2 c}$$

$$V = \frac{m}{p_{\text{воды}}} = \frac{m_n - m_1}{p_{\text{воды}}} = \frac{0,045 - 0,0275}{1} = 0,0175 \; \partial \text{м}^3$$

$$t = 180 \, \text{м} = 180 \cdot 60 = 10800 \; c$$

The data on water permeability of cotton fabric obtained by calculation are entered in Table 5.7 and a graph is plotted on their basis (Fig. 5.12).

Table 5.7.

Water resistance of excavated samples

linen	h\b	Time
0,0001632	0,0001006	60
0,0000896	0,0000607	120
0,0000602	0,0000405	180

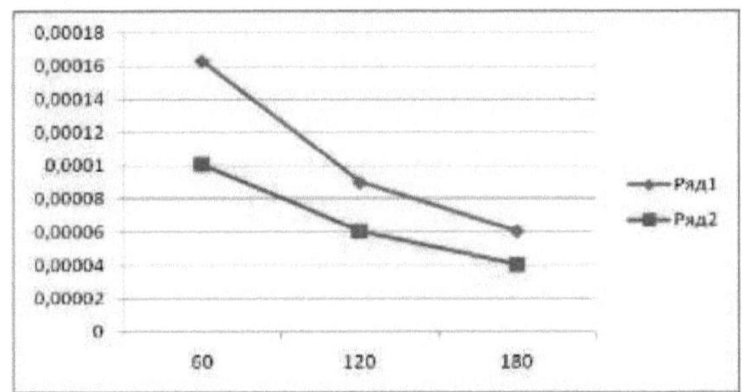

Fig. 5.12. Regularity of change of fabric water resistance depending on the exposure time: row 1 - for cotton yarn, row 2 - for linen yarn.

Physical properties of fabrics include: air permeability, water permeability, wear resistance, abrasion, etc. The requirements to the physical properties of fabrics are determined by their purpose and depend on the fibre composition, structure and finishing of fabrics. External friction against surrounding objects, causing wear of the material due to abrasion, occurs at the points of actual contact of the contacting surfaces. The nature of fibre fracture at the contact points is determined by both the structure of the material itself and the type of abrasion surface. There are two boundary types of fibre fracture: splitting into separate longitudinal structural elements caused by repeated impact of abrasive surfaces (fatigue wear) and micro-cutting caused by a single impact of contact surface protrusions. The cause of wear of fabrics is the impact on them of a complex set of different factors: mechanical, physicochemical and biological. Mechanical effects include abrasion and repeated stretching and bending, as well as compression and torsion; physico-chemical - the effect of light, atmosphere, moisture, temperature, sweat, detergents during washing and dry cleaning; biological - rotting processes caused by various microorganisms. Fabric abrasion resistance is one of the main indicators characterising its durability and depending on the structure of the fabric surface, as well as on the resistance of fibres or material to friction. The main indicators of the worked samples are presented in Table 5.8.

Main indicators of the worked out samples

№	Indicator name	Unit of measurement	flax		cotton	
1	Surface density	g/m²	568,2	565,6 (3)	679,7	681,1(3)
			566,3		680,9	
			562,4		682,8	
2	Air permeability	cm /cm ^sec³²	2,47	2,436	0,52	0,54
			2,42		0,62	
			2,42		0,48	
3	Water resistance	mm H2O	280	313, (3)	410	410
			320		420	
			340		400	

The mass of a fabric is expressed by a characteristic called surface density. The surface density of a fabric is an indicator characterising the mass of a unit area g/m^2 . This indicator depends on the thickness of main and weft yarns, fabric density and the nature of finishing. Thus, the surface density of wool fabric decreases after washing, boiling, bleaching and increases after felting, appretising, printing, etc. The surface density of textile fabrics is characterised by the thickness of main and weft yarns, fabric density and the nature of finishing. The surface density of textile materials varies within significant limits. It determines the purpose of the material. The surface density of textile materials is determined by weighing the materials or by calculation method. Water permeability is the ability of a fabric to absorb moisture and release water vapour emitted by the human body. Like hygroscopicity, these properties are most pronounced in fabrics made of natural fibres: linen, cotton, silk. Air permeability is a property that determines the ability of a fabric to let air pass through. The best "breathe" loose, thin fabrics made of natural fibres or fabrics produced by openwork weaves. This property is especially important for fabrics of summer assortment. Impregnation with an impregnating composition and application of film coatings considerably worsen the air permeability of fabrics. The comparative analysis of fabric parameters was carried out for the produced samples on the basis of the indicators, which are presented in Table 5.9.

A comparative analysis of tissue parameters.

Name	Unit of measurement	cotton			linen		
Surface density	g/m2	679,7	680,9	682,8	565,7	564,1	565,1
Water resistance	mm. water	410	420	400	280	320	340
erasure	sH	over 35,000			over 35,000		
Base density	yarn/dm	16			12		
Weft density	yarn/dm	10			8		
Linear density on the base	Tex	100x2			125x2		
Linear density in weft	Tex	120x2			170x2		

Classification of methods of weft insertion into the pharynx by means of insertion of weft insertion with a shuttle-gripper has been clarified. The existing systems of weft insertion by shuttle-gripper are complicated in design and require a number of additional mechanisms for tissue formation.

It is expedient to develop a new system of weft laying by shuttle-gripper on the basis of shuttle loom of AT type. Modernisation of the shuttle weaving machine is carried out by using the character of the movement of the reims with the installation of the mechanisms for feeding and withdrawal of the weft to the shuttle grip on the reims. The parameters of the shuttle-gripper are determined. Regularities of movement of the shuttle-gripper in the shed are obtained. The equations of tension of a sinker sliding on a plane, on a circle of stationary and mobile cylinders, taking into account rigidity of a sinker, radius, angle and coefficient of friction are received. It is expedient to use a movable cylinder as a shuttle catch. The values of the angle of friction of the sinker against the gripper depending on the position of the shuttle-gripper in the shed are determined. A stand and a methodology for determining the coefficient of friction of the weft against the shuttle grip depending on the shape, size and condition of the friction surfaces of the shuttle grip have been developed. Increasing the radius of friction of the yarn on the hook grip leads to an increase in the tension of the weft yarn. The coefficient of friction at rest is greater than the coefficient of friction in motion in all variants of rubbing surfaces thread-grip shuttle. The new fabric has a selvedge, the structure of which provides strength and prevents weaving of the extreme warp yarns and does not require additional mechanisms of selvedge formation. In terms of physical and mechanical properties the new fabric is not inferior to the standard fabric. The weft loss for the new fabric increases, which ranges from 5.6 to 9%, depending on the size of the weft loop in the selvedges. The fabric was designed for the production of fire hoses, as a result of which the necessary technological parameters for the production of the fabric were calculated. Our fabric should have water-resistant properties, which can be achieved with special impregnations. The fabric samples were tested for physical and mechanical properties in the laboratory of the certification centre, the characteristics of breaking load, resistance to abrasion, water resistance were studied. Our fabric has good tensile properties, breathability and water resistance.

LITERATURE

1 . HANDBOOK OF WEAVING Edited by S Adanur, Department of Textile Engineering, Auburn University, USA 440 pages 543 figures 68 tables 254 x 176mm hardback 2000.

2 .Sidorov Y. P., Rozanov A. F. "Analysis of works on automation of weft feeding of weaving machines abroad" Moscow, L. I. 1968

3 .Talavashek O., Svatyi V. "Needleless weaving machines" Moscow, L. I., 1985 4.Ormjord A., "Modern preparation and weaving equipment" Moscow, Legprombytizdat, 1987

5 F. M. Rozanov et al. "Technology of weaving" part 2, Moscow, L. I.,1967

6 Borodovsky M. S. et al. "Fundamentals of Weaving" part 2, Gizlegprom, 1947

7 .Urazbayev M. T. "Fundamentals of mechanics of weighted deformable flexible thread" Tashkent, 1951

8 .Kragelsky I. V. "Friction of fibrous substances" Moscow, Gizlegprom, 1941 9.Gordeev V. A. "Dynamics of mechanisms of tempering and warp tension" 10.Manukhin A. C. "Modernisation of mechanical weaving machine" Moscow, Gizlegprom, 1963

11.Meredith R. "Physical methods of research of textile materials" Moscow, Gizlegprom, 1963

12.Makarova T.A., Potapova L.V. "Textile Material Science" Moscow, 1986

13.Vlasov P. V. "Normalisation of weaving process" Moscow, L. I., 1982

14. http : //www.unfire01. ru/pozharnyj -magazin/rukava-pozhamye-napome. html

6 . GOST 29104.16-91. Technical fabrics. Method of determination water permeability.

7 . Ne'matov J.A., Rakhimkhodjaev S.S., scientific article "Fire hoses" Collection of scientific articles of master's students, Tashkent, TITLP, 2014.

8 . http : //www. abc 01. ru/rukava. php

9 . Eremina N.S., Koritsky K.I. Designing of cotton fabrics, Scientific and Research Works of the Central Scientific Research Institute of Textiles and Biotechnology, 1962.

10 . http://www.rukav.uz/

11 . GOST 9857-91 Cotton and mixed technical fabrics for rubber fabric sleeves. Technical conditions.

12 . http : //volbrok.ru/texnicheskie-tkani-i-rukavnye-filtry/

13 .

http://ru.wikipedia.org/wiki/%D0%9F%D0%BE%D0%B6%D0%B0%D1%80%D0%BD%D0%D 1%8B%D0%B9%D 1%80%D

1%83%D0%BA%D0%B0%D0% 14. http://chem21.info/info/604193/

15 . Ne'matov J.A., Rakhimkhodjaev S.S., scientific article "Linen fire hoses" Collection of scientific articles of master's students, Tashkent, TITLP, 2015.

16 . http://www.otkani.ru/property/phisicalproperty/6.html

17 . https://ru.wiktionary.org/wiki/%D0%B2%D0%BE%D0%B4%D0%BE%D1%8 3 %D0%BF%D0%BF%D0%BE%D 1%80%D0%BD%D0%BE%D1%81%D 1%82%D1

18 . http://www.otkani.ru/property/phisicalproperty/

19 . http://gearmix.ru/archives/4175

20 . http : //www. furfur. . me/furfur/all/futureclothing/169793-bezopasnost-podderzhka- kommunikatsii-peredacha-emotsiy-i-drugie-oblasti-primeneniya-umnyh-tkaney 21.Kolganova M.N., Taubkin S.I. "Application of chloroprene latex in the production of fire hoses", "Kauchuk i rubber", 1964.

22 Sukharev A.T. et al. Pressure hoses with braids from polyamide fibres, "Rubber and Rubber", 1963.

23 Koritskiy K.I. Use of chemical fibres in technical fabrics, NTO of light industry, "Light Industry", 1965.

24 .Orlik I.B. Application of synthetic fibres in the manufacture of fire hoses, CINTILegprom, Scientific information on flax-foam-jute industry, Series VI, No. 3, 1963.

25 .GOST 7877-75 Fire hoses pressure rubberised from synthetic threads. General technical conditions

26 Damyanov G.B. et al. Fabric structure and modern methods of its design. - M.: Light and Food Industry, 1984.

27 . Smirnov V.I., Theoretical studies of the structure of plain weave fabric, Rostekhizdat, 1960.

28 Gordeev V.A., Volkov P.V., Weaving M., L.I., 1984.

29 Kozireva Z.M. et al. Technical fabrics and their application, "Legkaya Industriya", 1965.

30 .Orlik I.B. Determination of linear percentage of filling of capron covers for pressure hoses, Scientific and Research Works of the Central Scientific Research Institute of LLV, vol. XXIII, publishing house "Light Industry", 1968.

31 .HANDBOOK OF YARN PRODUCTION Technology, science and economics P R Lord, NCSU, USA 504 pages 244 x 172mm hardback July 2003.

32 .Scientific article "Fabrics with special impregnations for mobile tents". Collection of scientific papers of the Military Academy of RUz, No.2, 2012.

33 .GOST 54873-2011 Nonwoven fabrics and products from them. Methods of determination of liquid transmission time.

34 GOST 29298-2005 Cotton and mixed household fabrics. General technical conditions.

35 .GOST 29298-92 Cotton and mixed household fabrics. General technical conditions.

36 GOST 20232-74 Cotton and mixed departmental fabrics. Standards of resistance to abrasion.

37 GOST 20359-74 Cotton and mixed departmental fabrics. General norms of air permeability.

38 Lund-Iversen B. Weaving weaves: Per. from Norv. - M.: Legprombytizdat, 1987.

39 Kutepov O.S. - Structure and design of fabrics. M., Legprodat, 1988.

40 .GOST 9857-91 Cotton and mixed technical fabrics for rubber fabric sleeves.

41 GOST 21790-76 Cotton and mixed fabrics for clothing. Technical conditions.

42 . http : //www. pojarnie-rukava. ru/.

43 . http: //museion. ru/1.8/tkan. htm

More
Books!

info@omniscriptum.com
www.omniscriptum.com
OMNIScriptum

Printed by Books on Demand GmbH, Norderstedt / Germany